The Letort Papers

REAL LEADERSHIP AND THE U.S. ARMY: OVERCOMING A FAILURE OF IMAGINATION TO CONDUCT ADAPTIVE WORK

John B. Richardson IV

December 2011

*****

Published by Books Express Publishing
Copyright © Books Express, 2012
ISBN 978-1-78039-668-2

Books Express publications are available from all good retail and online booksellers. For
publishing proposals and direct ordering please contact us at: info@books-express.com

CONTENTS

ILLUSTRATIONS

FOREWORD

To win on today's complex and competitive battlefield our military leaders have had to try to shed decades of organizational culture that emphasized control and stability as the solution to solving problem sets. Instead, today's leaders must be adaptive and agile in their analysis and development of innovative solutions to the complex challenges of the 21st century. Today's security environment requires men and women in uniform to think critically and be creative in developing new strategies and solutions. These skills will allow our military leaders to maintain the operational initiative against an enemy who is by nature adaptive and always evolving to overcome the tremendous advantage in technological and material overmatch of the United States and many of its allies.

This paper argues that the U.S. Army should continue its bold initiatives in its current Campaign of Learning and go even further. It should develop creative leaders who can exercise adaptive leadership with the capacity to provide learning environments within their organizations. Included in the paper is an analysis of adaptive challenges facing the Army. Specifically, the Army espouses the need for decentralized operations and operational adaptability, but the author argues that the Army culture is driven by control, stability, and risk aversion.

A case study provides a means for analyzing the complexity of organizational leadership in the contemporary security environment. The study presents a high-stakes problem set requiring an operational adaptation by a cavalry squadron in Baghdad, Iraq. This problematic reality triggers the struggle in finding a creative solution, as cultural norms serve as barriers against overturning accepted solutions that have proven successful in the past, even if they do not fit today's reality. The case highlights leaders who are constrained by assumptions and therefore suffer the consequences of failing to adapt quickly to a changed environment. Emphasizing the importance of reflection and a willingness to experiment and assume risk, the case study transitions to an example of a successful application of adaptive leadership and adaptive work performed by the organization.

The case study serves as a microcosm of the challenges facing the U.S. Army. The corresponding leadership framework presented can be used as a model for the Army as it attempts to move forward in its efforts to make adaptation an institutional imperative (Chapters 1 and 2). The paper presents a holistic approach to leadership, whereby the leader transcends being simply an authority figure and becomes instead a real leader who provides a safe and creative learning environment for the organization to tackle and solve adaptive challenges (Chapter 3). The paper concludes with a recommendation that Army leaders apply Harvard Professor Dean Williams's theory of leadership to the challenges confronting the Army's leader development process so as to improve its efforts to grow adaptive leaders (Chapter 4).

DOUGLAS C. LOVELACE, JR.
Director
Strategic Studies Institute

ABOUT THE AUTHOR

JOHN B. RICHARDSON IV assumed command of the 3d Cavalry Regiment at Fort Hood, Texas, in November 2011. He graduated from the U.S. Military Academy and was commissioned as a Second Lieutenant in Armor in 1991. He served as a tank platoon leader and company executive officer in Germany. After graduation from the Armor Officer Advance Course, he served as a Battalion S-1, Tank Company Commander, and Headquarters and Headquarters Company Commander. During this tour, he also deployed to Bosnia (IFOR) as the Command Liaison Officer to the Nordic-Polish Brigade. Colonel Richardson's previous assignments included duty as Company Tactical Officer at West Point, New York; 2nd Squadron Operations Officer (S3), 2d Armored Cavalry Regiment (Light), in Sadr City, Baghdad, Iraq; and Regimental Operations Officer (S3) at Fort Polk, Louisiana. He returned to Iraq in 2005 as the Aide-de-Camp to the Commanding General of Multi-National Security Transition Command-Iraq. In 2007 he assumed command of a cavalry squadron at Fort Riley, Kansas. During this 33-month command, the 750-soldier squadron task force served a 12-month deployment in northwest Baghdad in 2008-09. This deployment encompassed the final phase of "the Surge" campaign, the provincial elections, and a number of strategic transitions during the implementation of the security agreement between the U.S. Government and Government of Iraq. Colonel Richardson holds a master of science degree in counseling and leader development from Long Island University; has attended the U.S. Army Command and General Staff College; and was selected as a National Security Fellow for a

1-year Senior Service College Fellowship in leadership and management at Harvard's Kennedy School of Government.

SUMMARY

Former Army Chief of Staff General Martin E. Dempsey has highlighted "failure of imagination" as a major obstacle in an organization's ability to learn, adapt, and find solutions to complex problems. As a former Commanding General of the Army's Training and Doctrine Command (TRADOC), General Dempsey led the redesign of the Army's conceptual foundation. He and other Army officials, reflecting on the previous decade's conflicts, aggressively instituted a campaign of learning, which TRADOC describes as "a broad set of initiatives designed to produce an Army capable of rapidly adapting to defeat unforeseen threats."

This paper argues that the U.S. Army should continue its bold initiatives and go even further. It should develop creative leaders who can exercise adaptive leadership with the capacity to provide learning environments within their organizations. Included in the paper is an analysis of adaptive challenges facing the Army. Specifically, the Army espouses the need for decentralized operations and operational adaptability, but the Army culture is driven by control, stability, and risk aversion. The author provides a recommended solution for overcoming this disconnect and achieving adaptive leadership through the application of a leadership framework provided by Dean Williams of Harvard's Kennedy School. The focus of "real leadership" as presented by Williams is not to get others to follow, but rather is directed toward getting people to confront reality and change their values, habits, practices, and priorities to deal with the *real* threat or opportunity the group faces. It is through this more holistic approach to leadership that the

Army can maximize the adaptations in its campaign of learning and develop leaders with the capacity and skills to foster learning environments within their organizations.

In many cases in Iraq and Afghanistan, however, the Army is getting it right. Junior leaders, routinely exercising adaptive leadership, demonstrate numerous examples of how to overcome intractable challenges in changing and ambiguous environments. In these cases, commanders are leading their organizations through complex realities by assuming necessary risk and seizing the initiative from our adversaries. These examples of a more creative approach to problem-solving based on mutual trust inside of the organizations ensures that these units achieve and maintain momentum over the enemy, which is necessary to progress successfully in the fight against adaptive opponents in the current operational environment.

Unfortunately, the discretion and flexibility granted to many leaders in these theaters of war is not how business is done on a day-to-day basis throughout the U.S. Army. What is at stake if we do not incorporate these battlefield lessons into the organization's DNA will be a lower likelihood that the Army will produce the quantity and quality of leaders who are creative, imaginative, and innovative, and can lead learning organizations on today's competitive battlefield. It is the risk of operating at 80-percent effectiveness as an organization when 90 percent or better can be achieved with cultural alignment between what the Army says is important and what it actually rewards as success through professional advancement. The Army is at a pivotal juncture as it attempts to transition over the next decade from war in two major theaters back to a traditional garrison routine in the context of persistent global conflict. Its ability to institute an organizational change in culture that can produce operational adaptability is critical to successfully moving the Army toward making adaptive leadership the way it does business.

This paper begins with a case study that provides an example of a real-world adaptive challenge during a cavalry squadron's recent deployment to Baghdad in support of Operation IRAQI FREEDOM. The case study accentuates the challenges of organizational leadership in complex environments and the consequences of failing to adapt fast enough on today's battlefields. Ultimately, the unit makes in-stride adaptations, demonstrating how the exercise of adaptive leadership must start with a reflective diagnosis and an accurate understanding of the adaptive challenge. It highlights the role of the leader in presenting the reality of the changed conditions to the organization, and then providing a learning environment based on trust and empowerment to allow the group to develop adaptive solutions. Simply put, the case study confirms the requirement for operational adaptability to become "how we do business" in the security environment of the 21st century.

Following the case study, evidence is presented that suggests there is a lack of congruency between the Army's espoused values (what we say we should do), and the Army's basic underlying assumptions (what we actually do), which can cause a trust deficit and produce an organizational culture that is not conducive to the development of adaptive leaders. A review of the espoused values from emerging Army doctrine, focused on decentralized execution and operational adaptability, is contrasted with the conclusions of a study conducted by Dr. James Pierce at the Strategic Studies Institute of the U.S. Army War College. Dr. Pierce's findings suggest a lack of congruency between espoused values and actual practice. His study brings into question whether the Army's organizational culture is one that encourages the kind of imagination necessary to achieve operational adaptability. This cultural disconnect is the primary barrier confronting the Army's campaign of learning.

To overcome this divergence in the Army's organizational culture, the author presents Williams's framework for leadership as a way to help close the gap between espoused values and basic underlying assumptions. Williams posits a theory of leadership from which the U.S. Army can glean critical insights in its quest to change its culture and achieve operational adaptability. Williams postulates that traditional notions of leadership are inadequate for today's challenges — that they do not address the complexities and diversity of the problems, threats, and opportunities that groups and institutions must confront in today's globalized and complex world — if these groups and institutions expect to progress. Traditional notions of leadership unduly emphasize the role of the leader in providing a vision or "showing the way," while leading in primarily straightforward environments. In contrast, Williams addresses the demanding task of mobilizing people to confront their predicament and solve their most pressing problems.

RECOMMENDATIONS

- Eliminate completely the current officer evaluation system and corresponding promotion system. Replace the system with one that identifies, develops, and rewards adaptability, creativity, entrepreneurial behavior, and prudent risk-taking.
- Initiate a 360-degree developmental feedback process for all noncommissioned officers (NCOs) and commissioned officers to identify bad (counterfeit) leaders and develop all leaders in the spirit of the professional military ethic (PME).
- Expand the Army's definition of leadership beyond the present scope of providing purpose, direction, and motivation so as to include as well mobilizing people to confront and address problematic realities to accomplish adaptive work.
- Expand the Army's *Leader Core Competencies* to include the following additional competency: "**One who mobilizes**. Navigates organizations through adaptive challenges by confronting people with the reality of changed conditions and provides a learning environment for the group to discover and develop adaptive solutions" (see Appendix III).

METHODOLOGY

A case study provides a means for analyzing complex organizational leadership in the contemporary security environment. It presents a high-stakes problem set requiring an operational adaptation by a cavalry squadron in Baghdad, Iraq. This problematic reality triggers a struggle in finding a creative solution, as cultural norms serve as barriers against overturning accepted solutions that have proven successful in the past, even if they do not fit the reality of the present (failure of imagination). The story highlights leaders who are constrained by assumptions and, as a result, fail to adapt quickly to a changed environment. Emphasizing the importance of reflection and a willingness to experiment and assume risk, the case study transitions to an example of a successful application of adaptive leadership and adaptive work performed by the organization.

The case study serves as a microcosm of the challenges facing the U.S. Army. The corresponding leadership framework presented can be used as a model for the Army as it attempts to move forward in its effort to make adaptation an institutional imperative, as shown in Chapters 1 and 2. This paper presents a holistic approach to leadership whereby the leader transcends being simply an authority figure and becomes instead a real leader who provides a safe and creative learning environment for the organization to tackle and solve adaptive challenges, as discussed in Chapter 3. In Chapter 4, the paper concludes with a recommendation that Army leaders apply Williams's theory to the challenges confronting the Army's leader development process so as to improve its efforts to grow adaptive leaders.

> To increase the chances of success, real leadership must be approached as an interactive art. It is an art in that it requires creativity and imagination, rather than a singular set of well-honed practices; and it is interactive in that one must be willing to "dance" with the reality of the context so that the best solutions can

emerge…real leadership requires a capacity to improvise, be imaginative, and make ongoing corrections according to the specific challenge the people face, the discoveries of the group as they tackle the challenge, and the shifting dynamics of the context. Therefore, strong diagnostic skills and considerable flexibility in one's intervention style are essential if one is to lead effectively in multiple contexts on multiple challenges.[5]

— Dean Williams

ENDNOTES - INTRODUCTION

1. Rollo May, *The Courage to Create*, London, UK: Norton, 1975), p. 60.

2. Plato, *The Republic*, New York: Oxford University Press, 1945, pp. 232-233.

3. Martin E. Dempsey, "Driving Change Through a Campaign of Learning," *Army Magazine*, Vol. 60, No. 10, October 2010, p. 66.

4. *Ibid.*

5. Dean Williams, *Real Leadership: Helping People and Organizations Face Their Toughest Challenges*, San Francisco, CA: Berrett-Koehler Publishers, 2005, p. xiii.

CASE STUDY

IRAQ 2008-09:
OVERCOMING A FAILURE OF IMAGINATION
TO CONDUCT ADAPTIVE WORK

Just as our adversaries adapt and develop new tactics, techniques, and procedures, we too must be nimble and creative.[1]

— *National Defense Strategy* 2008

The following case study is a firsthand account of the author's experience while serving as a squadron commanding officer (SCO — pronounced Sko) of an armored wheeled and mechanized cavalry squadron operating in Baghdad, Iraq. The study provides relevant parallels with the three key elements of the adaptive challenges facing the Army: (1) emerging doctrine on adaptation, (2) hindered by a culture of control, stability, and risk aversion, which can be (3) overcome by a leadership framework based on mutual trust, experimentation, and risk-taking. Challenging the status quo should become a core attribute of the Army culture. Continuously challenging all aspects of the "way we do business around here" should become an incontrovertible mandate that is simply understood as a critical component of a healthy self-reflexive profession, and is therefore encouraged and rewarded.

BACKGROUND

In the fall of 2008, 2nd Brigade of the 1st (U.S.) Infantry Division deployed to Iraq and was attached to the Multi-National Division-Baghdad (MND-B), where it would serve under the command of the 4th Infantry Division (3 months) and later the 1st Cavalry Division (9 months) in the MND-B battlespace. From late 2008 into the spring and summer of 2009, the brigade's cavalry squadron, a subordinate battalion-size organization of 750 Soldiers commanded by a lieutenant colonel, was faced with an adaptive challenge. The enemy in the squadron's area of operation (AO) had adapted to the U.S. Army's successful technological solutions to counter the improvised explosive device (IED) threat and, as a result, the enemy had changed its primary weapon system from the IED to the RKG-3 high explosive anti-tank (HEAT) hand grenade. Along with this new weapon came new enemy tactics. Instead of remotely blowing up a roadside bomb from a safe and undetectable location, the RKG-3 was employed in a face-to-face close combat attack. The enemy would suddenly emerge from alleyways within hand grenade throwing distance (10-30 meters) of the passing vehicle and then execute a classic hit-and-run irregular ambush.

This case study is a story of a squadron struggling with a changed environment and of its authority figure, the SCO, whose initial lack of imagination resulted in what would be described by Dean Williams as **counterfeit leadership**.[2] This prevented him from exercising real leadership until he stepped back from the problem, actively listening and sensing the environment, reflecting, then properly diagnosing the problem, and determining the need to empower the unit to conduct adaptive learning. It is the journey that one unit made to face the reality of a changed environment, to understand a complex problem, to reframe it, and then through the exercise of leadership, to eventually induce the squadron to conduct the necessary

adaptive work[3] to defeat the threat. Ultimately, as a result of successful tactical engagements based on this adaptive work, the unit not only protected itself from enemy tactics, but the success of exercising leadership by mobilizing the group to achieve adaptive solutions *created a new culture of operational adaptability,* which allowed the squadron to go beyond just defeating a tactical threat. This exercise of leadership helped change the organization's culture and redefined the unit's approach to all future adaptive challenges, ultimately allowing the organization to destroy an entire insurgent network with experimental, creative, and innovative combined operations by, with, and through the 54th Iraqi Army Brigade in northwest Baghdad.

UNDETECTED CHANGE IN THE ENVIRONMENT

This journey began in September 2008, while 5th Squadron, 4th Cavalry Regiment (5-4 Cavalry), conducted a relief-in-place (RIP) with 1st Squadron, 75th Cavalry Regiment (1-75 Cavalry), in northwest Baghdad. After 5 years of unit rotational transitions, units had become very adept at transferring authority for an AO as new units replaced outgoing units. 1-75 Cavalry was completing a 15-month tour as part of the Surge Campaign. Its tour was highlighted by the traditional combat threats of Operation IRAQI FREEDOM (OIF): snipers and IEDs. In April and May 2008, 1-75 Cavalry participated in high-intensity close combat during a Shia uprising in the Shula neighborhood. After they soundly defeated this threat, the remainder of their tour was focused on consolidating the gains of the clear/hold stages articulated in the updated U.S. counterinsurgency (COIN) doctrine of clear/hold/build. The 1-75 Cavalry then began to rebuild their area under the umbrella of the security they and their Iraqi Security Force (ISF) partners had established during the Surge. The 1-75 Cavalry's AO encompassed Shia and Sunni neighborhoods, with an ethnic fault line presenting most of their challenges. The 1-75 Cavalry's positive relationship with the Sons of Iraq (SoI) elements (former Sunni insurgents turned allies in the fight against al-Qaeda in Iraq) meant that they faced minimal anti-Coalition activity in their Sunni neighborhoods. Violence was down by 80 percent across Baghdad, and the operational results of the Surge seemed to indicate a turning point in the conflict. The achievement of irreversible momentum was growing legs and becoming part of the narrative.

As 5-4 Cavalry conducted the 3-week RIP transition with 1-75 Cavalry in October 2008, new internal brigade boundaries were adjusted, and most of the Shia population centers were shifted from 5-4 Cavalry's AO into the AO of its sister battalion, 1-18 Infantry, on their flank. In essence, 5-4 Cavalry now had primarily a Sunni AO, which had been relatively quiet over the last 9 months. The U.S. Forces (USF) enjoyed a positive and productive relationship with both the Iraqi Security Forces (ISF) in the AO and, more importantly for local security, a strong relationship with three SoI battalions in the neighborhoods. It appeared that for 5-4 Cavalry's tour, combat operations would be minimal, and that fostering greater partnership with the ISF and furthering the development of the ISF would be the primary focus for the next 12 months. The squadron would continue the "build" activities of the clear/hold/build strategy and prepare for the eventual passing of security responsibilities to the ISF by the end of the tour.

The first indicator that the tour would not be as assumed came on the night of October 21, 2008, during the RIP transition with 1-75 Cavalry. The troop commander (captain) of Troop A, 1-75 Cavalry and the troop commander (captain) of Troop B, 5-4 Cavalry, rolled out of the squadron Joint Security Station (JSS) — small USF and ISF joint outposts in the neighborhoods — to conduct a routine patrol to familiarize the Troop B commander with his soon-to-be AO. As the four-vehicle patrol left the safety of the JSS and moved into the neighborhood via Route Cecil, an explosion alerted the JSS of enemy contact just outside the gate. Initial assessment was an IED attack on the passing patrol — the first in months against 1-75 Cavalry in this Sunni neighborhood. Further investigation by a U.S. Explosive Ordinance Detachment (EOD) determined the explosion was *not* caused by an IED, but rather by an RKG-3 hand grenade that partially malfunctioned, causing minimal damage to the up-armored HMMWV (Humvee). No one fully appreciated the significance of the appearance of this new weapon on the battlefield. It was considered an anomaly and, because no one was injured, was forgotten almost as fast as it happened. The units' transition process continued as planned.

Two weeks later during the final hours of the RIP, the two chains of command were meeting in the squadron headquarters (HQ) for the final out-brief from 1-75 Cavalry to 5-4 Cavalry to ensure that all key information and intelligence, and all current tactics, techniques, and procedures (TTPs), were shared; that any previously pending issues had been resolved; and that 5-4 Cavalry was prepared and felt comfortable assuming control for the 1-75 Cavalry AO. As the meeting was coming to a close, a large explosion sounded.

About two kilometers (km) south of the JSS, a 1-75 Cavalry patrol coming to pick up its troop commander from the meeting was hit. A runner from the squadron tactical operation center (TOC) came into the conference room. He reported there were casualties this time. The meeting broke up, and Soldiers went into the battle drills for reacting to such an event. The casualties, which were severe, had been evacuated to the Combat Surgical Hospital (CSH) across the city in the Green Zone. The two squadron commanders (SCOs) and command sergeants major (CSMs) moved immediately to the CSH. A platoon sergeant, on his third combat tour and completing what would have been his last patrol of this 15-month tour, had lost his arm in the attack. The weapon, which also destroyed the vehicle and wounded all its occupants, was an RKG-3. The 5-4 Cavalry SCO directed the staff to produce an information brief on this new weapon the following morning.

The RKG-3 is an anti-tank hand grenade. It has an armor-piercing copper shaped-charge that can penetrate 9.5 inches of armored steel. It is thrown by using a wooden handle (it looks like a German "potato masher" grenade from World War I), and explodes on impact. This creates a small quarter-size hole of penetration, producing shrapnel and spalling inside the vehicle, with the molten copper dart itself destroying anything in its path. Originally designed by the Soviet Union and carried by Warsaw Pact infantry to be used in close combat with North Atlantic Treaty Organization (NATO) tanks, its proliferation around the world makes the RKG-3 a cheap and easily accessible weapon for insurgents. For 5-4 Cavalry, it was an enemy adaptation that had allowed the insurgents to seize the initiative.[4]

FAILURE OF IMAGINATION: COUNTERFEIT LEADERSHIP

The 5-4 Cavalry officially assumed responsibility of the AO on November 9, 2008, and went to work to accomplish its mission: "To train and support partnership-unit Iraqi Security Forces to secure the population . . . to enable continued progress toward achieving sustainable security and ensure continued development of Iraqi civil capacity." Then, on December 1, 2008, another RKG-3 attack occurred in which the 5-4 Cavalry experienced its first casualties. One Soldier lost his leg, and a dedicated Department of Defense (DoD) civilian was killed. There was disequilibrium throughout the system, as frustration was running high. The commander felt compelled to prevent another attack in order to protect his troops. As the Soldiers looked to their commander for solutions, the SCO prepared to provide those answers. He had trained for 17 years to prepare himself to be a leader in this situation, to use good judgment and make decisive, ethical, and tactically sound decisions in a time of crisis. He was *the* commander, the premier **authority**[5] in the organization under attack, and in this profession. That meant he was responsible for everything the unit did and failed to do, and at that moment the unit was failing to protect itself. In turn, as the SCO prepared to provide answers and solutions, he too looked to authority for solutions. He looked to the commanders of the SoI—the Iraq informal authority in the neighborhoods—for answers. Since these were Sunni attacks in Sunni neighborhoods and USF were paying the SoI commanders to prevent this kind of activity, the SCO planned to hold them accountable for results.

The SCO reviewed what he knew and provided motivational speeches about greater vigilance and enforcing standards of current TTPs to inspire and calm. He gave sincere answers and provided well-intentioned **technical solutions**[6] based on previous personal combat actions and years of experience, all typical and standard behavior for a commander seeking to lead his organization through challenging and dangerous times. In retrospect, however, his performance evinced a failure to exercise real leadership and a complete failure of imagination.

REFLECTION AND ANALYSIS: BENEFITS AND LIMITATIONS OF AUTHORITY

One of the challenges in leadership is understanding and then managing the use of authority. Authority is power granted to perform a service. Formal authority in the military is manifested when a commander commits to meeting a set of explicit expectations, as defined by the commander's job description and the profession's standards. The Army is command-centric. Stated another way, it places tremendous responsibility on authority to get things done. This paradigm creates an expectation that the commander must provide answers; that the leader embodies the all-knowing; and that, as a result, authority in the military can be inappropriately substituted for leadership. When a problem is too complex for the leader to solve alone, a culture of authority-centric leadership places constraints on finding adaptive solutions. Conversely, if a leader uses authority to set conditions for a group to conduct experiments, which results in discovery learning, then authority becomes a combat multiplier. In this paradigm, authority is used to create space for the group to do adaptive work and find solutions to complex problems.

Within days the SoI leaders provided "actionable intelligence." They provided the typical technical solution the unit sought; paid informants came forward who could point out where the alleged perpetrators bedded down. The squadron acted decisively, conducting midnight raids and arresting accused suspects. The raids brought back a sense of equilibrium and curbed the sense of helplessness for the unit, but the ensuing tactical questioning and interrogations by the ISF yielded no proof that the unit had picked up the right people.

Although equilibrium and morale were temporarily restored, the SCO's intuition was that they had not solved the real problem. In an attempt to find a solution, the SCO reverted to another time-tested technical response: he would bring in the "big dog." He would ask the brigade commander (higher HQ authority) to come to the next neighborhood council meeting and confront and challenge the local civil leadership. The brigade commander met with local Iraqi leaders and demanded that the attacks on 5-4 Cavalry stop or the money he controlled for infrastructure projects would cease to flow into their communities. Pressure was also applied to the partner ISF units, but they were as clueless as the USF as to the source of this new threat. They even denied that the RKG-3 was the weapon being used (despite the USF leaders holding the safety pin and other RKG-3 components recovered from the attack sites and the enemy posting propaganda videos of the attacks on the internet).

Two weeks of quiet gave the unit a sense that the problem had been solved. They had looked to their authority figures—the Soldiers to the SCO and their troop commanders, the SCO to the SoI leaders (informal authority) and the brigade commander (formal authority); the unit had used the tools of the past to apply technical solutions to the problem—increased vigilance, strict adherence to standard operating procedures and TTPs, raids and apprehension of suspects—and threatened to cut off money for local infrastructure projects. The problem seemed to be resolved. Then on December 14, 2008, their equilibrium was once again disrupted as another patrol was hit by an RKG-3 on Route Cecil. The gunner survived, although he was severely wounded. The gunner and his crew were doing all the right things, using all the time-proven TTPs the unit had used during its train-up and certification for deployment. Why, then, were he and his crew members tactically defeated?

FACING REALITY AND DIAGNOSING THE REAL PROBLEM

The unit was being defeated because during their train-up for deployment the threat (which had become the routine problem on the battlefield of Iraq) was the IED. Time-tested, counter-IED tactics and procedures had proven to be effective in the past for solving the problem (force protection in Iraq) and had become an accepted way that USF did business. The IED is what they trained to defeat, and they were trained well. There is no doubt that if this crew were faced with an IED, the results would have been different. But the enemy had adapted; the environment had changed.

Five years of aggressive U.S. counter-IED technology and tactics had rendered the Sunni IED practically ineffective against the USF and its improved armor and technological countermeasures. If the enemy wanted to remain credible in the eyes of the population and relevant in the internal struggle for power in the Iraqi political system, he would need to adapt. The Sunni insurgent group, the *1920's Revolutionary Brigade*, in northwest Baghdad had done exactly that. They changed TTPs to adapt to the USF strengths and became a credible irregular military threat once again. They stopped risking their lives emplacing IEDs, whose lethal effects had so dramatically decreased, and picked up the RKG-3 hand grenade in its stead.

The SCO had to facilitate a solution to this tactical problem. The problem was complex, and the solution, if there was one, seemed hidden. The previous solutions provided by the SCO made people feel good, like they were doing something to solve the problem, but as far as truly solving the real problem was concerned, the familiar solutions did nothing. The SCO was providing moral support, setting the personal example by leading patrols and not asking his Soldiers to do anything he would not do himself, but he was not exercising leadership in the face of a complex problem that had no technical solution that could be pulled out of an already established repertoire. At best, it was counterfeit leadership, a bandage to make the unit's members feel as though they were on the offensive, as if they had the initiative by chasing false tasks, and as if they were winning. In fact, the opposite was true, and it was now the leader's responsibility to present this reality to the organization and get down to the real work at hand.

REFLECTION AND ANALYSIS:
UNDERSTANDING THE SYSTEM

One of the challenges in leadership is understanding and then managing the system in which the leader is immersed. A military organization is a system of systems, made up of different factions, each with its own loyalties, values, and relationships within the system. These factions can place obstacles in the path of progress and thwart the exercise of leadership, because their conflicting loyalties and values create agendas counter to moving the group toward an adaptive solution. These obstacles could come in the form of higher HQ using authority to discourage experimentation because of the risk and fear that failure could reflect poorly on them. It can be a faction inside the unit that is clinging to past assumptions or a faction that does not see the benefit of offensive action and just wants to survive and get everyone home alive.

These systemic dynamics, reinforced by a hierarchical organization prone to risk aversion, make it difficult for a military leader to exercise leadership. Navigating the system and being able to step out and put oneself at risk requires a leader to shape conditions within the system so that imaginative and creative problem-solving can take place. Forming partnerships and co-opting factions within the system are critical in setting the conditions for experimentation and risk-taking. This may mean getting buy-in from the chain of command to provide top-cover in the event of a failed experiment or educating those in factions below that the risk and experimentation are actually aligned with their current loyalties, practices, and beliefs. In this case it means selling the concept to resistant factions that innovative offensive solutions are the best way to accomplish the mission while increasing the chances of bringing everyone home alive. In this scenario, the commander brings together elements of multiple factions that share similar goals and empowers them to develop solutions. By co-opting multiple factions, when it is time to move the group toward accepting required adaptation, the leader sets the conditions for removing enough institutional inertia by building consensus and allowing the group to progress.

INTERVENTION: PUTTING THE WORK ON THE GROUP

The SCO decided to bring a larger group of diverse leaders together, present them with a diagnosis of the problem, and put the group to work to come up with adaptive solutions to their challenge. The SCO brought in 12 members of the command, ranging from staff sergeant to major. Rank and authority played little, if any, part in the decision on whom to bring to what would become known as the RKG-3 Defeat Working Group. The criterion for selection was based on the SCO's personal observation of these leaders over the previous 20 months. The selection criteria included leaders who had developed innovative training in the past, leaders who showed a propensity for taking prudent risk, leaders who had invented new tactics or new equipment configurations, and leaders who demonstrated an ability to transfer knowledge from one scenario to another and were skilled at imbuing it in others as well.

The diagnosis revealed that the unit had an enemy who had changed TTPs and weapon systems to adapt and overcome USF strengths, which were improved armor protection and counter-IED technology. The enemy adapted by going low-tech—running from dark alleys and throwing armor-piercing hand grenades. No high-tech jammer was going to stop an individual from running and throwing a grenade. Though it was 1950s Soviet armor-piercing technology, it was sophisticated enough to penetrate the up-armored HMMWVs and mine-resistant, ambush-protected (MRAP) armored fighting vehicles that were employed to counter the IED threat in Iraq. Furthermore, the USF counter-IED mentality was defensive in nature. The mindset was: jam signals that trigger the IED, wrap armor around yourself, and hope the IED does not go off, or if it does, that it is misdirected or the armor prevents catastrophic penetration.

After presenting this reality, the SCO provided the following guidance to the working group:

> Don't fight the 'last war' or your last deployment, take the gloves off, anything goes, think outside the box, question assumptions, be creative, use your imagination. . . . I want to hear any and every idea you have. Weigh risk, but do not let risk prevent you from presenting an idea . . . it may not be as risky as driving down Route Cecil under current conditions.[7]

The SCO left no one in charge and left the conference room, shutting them in alone.

The SCO returned 1 1/2 hours later and joined the discussion. The group had formed its own dynamics; still, no one was in charge, and the commander reintegrated into the group as a contributing member and not as "the SCO." The intent of leaving the room had been to take authority out of the equation. If the SCO had stayed, the group, by the very nature of group dynamics and Army culture, would have looked to him for solutions, and even if he put the work back on them, answers would be generated to satisfy the authority and not to attack the problem with an open mind. When he returned, the group had mobilized to do work, and the power and momentum of the group were not derailed by the reappearance of authority. The group knew it was on the path to adaptive solution(s). The SCO joined the discussion as a group member, not as the leader, and as a group it developed a framework for how to overcome its adaptive challenge.

ADAPTIVE WORK AND OVERCOMING BARRIERS

From this session, the RKG-3 holistic defeat approach was born: attacking the thrower, attacking the network, and co-opting the population. The immediate focus was how to change, modify, and adapt the TTPs to attack the thrower. This was a critical force-protection issue; the unit had to protect itself with new tactics and because it believed that success in this critical task would lead to success in the other two.[8] The rallying cry which sprang from the meeting was: "Be the hunter, not the hunted!" This problem was very complex and, as a result, the adaptive solutions were far-reaching. For the purpose of this case study, covering all of or listing the actual experimental techniques and innovative solutions in their entirety is beyond the scope of this case study, but shared below are a few of the proposed solutions and insights that are indicative of the adaptive mentality of the group.

First, the enemy always attacked the last vehicle in the patrol. It was the most vulnerable, because the other vehicles had passed the point of attack, allowing the attackers time and space to make their escape. Providing protection for the last vehicle was thus critical and became the focus of the new tactics. Protection from IEDs meant more armor protection for the gunner who sits exposed through a hole in the roof of the vehicle. This increased armor, while it protects the gunner from IED shrapnel, also drastically decreases visibility and severely limits the gunner's ability to maneuver his weapon to react to the RKG-3 thrower.

Target acquisition was identified as the single most important factor in preventing an RKG-3 ambush, an issue not relevant in the IED fight where the trigger man is likely out of sight and provides no sign to identify himself as the target. In an RKG-3 ambush, the gunner has a split second to identify the attacker, who will be hiding in a crowd or down a dark alley; assess hostile intent; bring his weapon into position; take aim; and fire from a moving vehicle before the attacker can make a baseball-style throw and run—a seemingly impossible task. The recommendation from the group: take a welding torch to the armor plates surrounding the gunner. Cut the plates away and expose the gunner so he can see and maneuver freely. This was considered radical, despite being common sense. The commander was all for it. The SCO's Personal Security Detachment (PSD) was the first to implement it. Figure CS-1 is a slide from 5-4 Cavalry's TTP briefing demonstrating the "before and after" look of the turret based on the recommended modifications.[9]

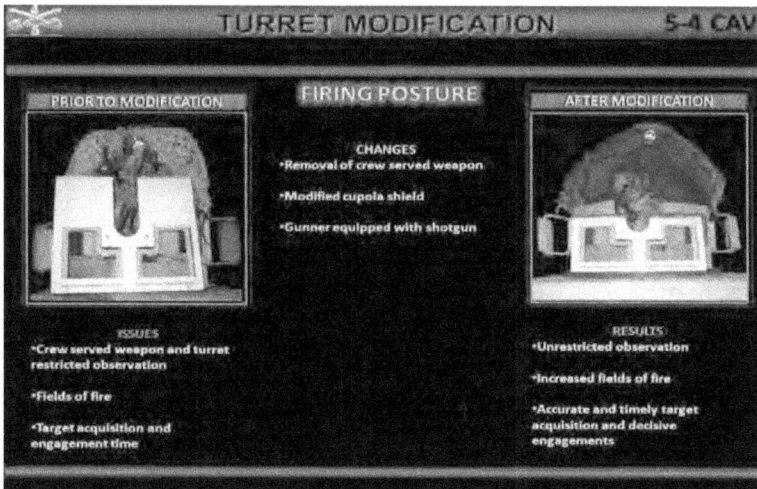

Figure CS-1. "Before and After" Recommended Turret Modifications.

This solution met immediate resistance within the squadron as a whole. Those who did not have the benefit of being part of the working group had not come to grips with the reality of the current threat environment. Noncommissioned officers (NCOs) with multiple tours in Iraq (some saved by armor protection from IED attacks in previous tours) protested that these adaptations assumed unnecessary risk in exposing the gunner to an IED explosion. Their lack of understanding limited their innovative spirit and prevented them from seeing the value of experimenting.

On one particular day, the SCO became engaged in a heated discussion with a passionate platoon sergeant, who refused to allow the welder to cut the armor plates from in front of his gunner. Asked to explain why, the platoon sergeant clung to past successful assumptions that the plates were there to protect the gunner from shrapnel from an IED and the sergeant cited a number of examples from his tour in Iraq in 2005, when the plates saved the gunner's life. The platoon sergeant was then asked: When was the last IED attack in the 5-4 Cavalry AO? In fact, there had been no IEDs in 2008 in the AO being discussed. Then the sergeant was asked how many RKG-3 attacks there had been. By the end, the SCO was able to help him see the reality of the current situation, and the welder went to work.

This encounter brought to the forefront the challenge that leaders confront when having to ensure the *whole* organization faces the reality of the current situation. This realization meant the smaller cohort of leaders from the working group had the responsibility to go back to their organizations and mobilize the larger group. The SCO and CSM intervened by going on the circuit—a "world tour" explaining the problem, laying out the diagnosis, and leading the group in the learning process. They owed it to the Soldiers and the junior leaders of the organization to explain the "why" so they understood and could have ownership for the experimental procedures and other adaptations. They met with each platoon (18 total) separately and explained the adaptive challenge and the adaptive work to be done. The concluding slide of the briefing used in the intervention with the platoons is shown in Figure CS-2.[10]

"We must remain a learning organization and we must adapt if we want to seize the initiative back from the enemy with RKG-3s...the enemy is in our OODA loop and we are reacting to him, often times in the form of CASEVAC. We must get into his decision making cycle and get ahead of him with new TTPs, new/modified equipment, and quality IPB. We must adapt and react faster than he can, take the initiative back...study, learn, adapt, and attack."

SCO

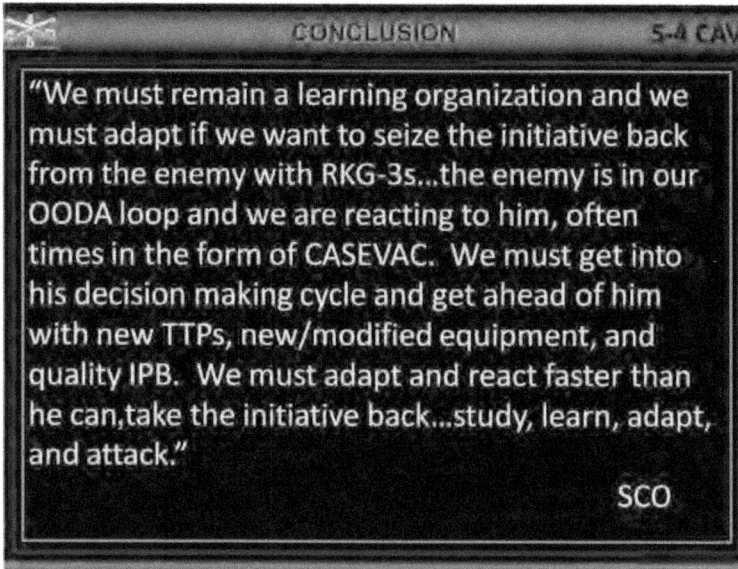

Figure CS-2. Platoon Intervention Briefing.[11]

These platoon briefings went a long way in helping the unit move forward. Despite these efforts, 10 to 20 percent of the organization refused to face the reality. Despite future success and validation of the adaptations, this small group of holdouts never saw the value in the innovations, experimentation, and ultimately the feasibility of the adaptive work. This fact further accentuates the challenge of leading people through the learning process when they hold threatened strong beliefs and trusted practices.

Another innovative solution, also part of solving the target-acquisition issue, was taking the crew-served weapon off the turret and arming the gunner with a shotgun instead. The mounted crew-served weapon was usually a .50 caliber machine gun, meant for engagements 500-1,500 meters away, or an M240 machine gun, likewise a weapon meant for a high rate of fire and longer-range engagements. Both can produce major collateral damage in an urban environment, with neither weapon conducive to rapid, precision engagements inside 50 meters. The working group challenged the assumption that a crew-served weapon was necessary at all. The fact that it impeded the vision and maneuverability of the gunner was further reason to challenge its presence. The fact was at this point in the campaign, Iraq 2009, that the unit was not going to face a threat such that every vehicle in a patrol would need that much firepower. The nature of the real threat, the 1-second RKG-3 engagement, would not be won by an unwieldy machine gun but rather by a quick-fire area blast of a shotgun, disrupting or unnerving the thrower and regaining the initiative.

By evening the odds, the movement to contact[12] had become a meeting engagement[13] instead of an ambush.[14] To lower the risk of collateral damage in this close combat engagement, where noncombatants would likely be caught in the crossfire, nonlethal crowd control shotgun rounds were used. This was because the throwers often would hide in crowds of innocent bystanders to camouflage themselves and use the civilians as shields for their escape, knowing the U.S. gunner would not likely fire into a crowd with lethal force, even in self-defense.

The experimental concept was this: Quickly identify the hostile intent, **get the first shot off** with an area fire blast of a shotgun and prevent or disrupt the throw. The crowd goes down, initially angry, wondering why they were just peppered by a crowd-control shotgun blast. The would-be throwers on the ground know exactly why the crowd was blasted, and they will be up and on their feet immediately to escape. By this point, however, based on new innovative battle drills, the patrol has stopped and dismounted scouts who, with precision M-4 carbines, can take aimed lethal shots at the fleeing ambushers. The crowd now realizes the rest of the story and is appreciative that USF, in self-defense, did not engage the embedded enemy with immediate lethal fire.

The abandonment of the crew-served weapons so ingrained in the mindset as "how we do business" and the introduction of nonlethal rounds in combat induced high levels of discomfort and resistance while the group struggled with the adaptive work. The squadron drastically modified equipment, changed tactics, drilled and rehearsed new actions-on-contact battle drills — all experimental in nature, all requiring risk, to adapt to the new enemy threat of the menacing RKG-3. On February 5, 2009, along Route Cecil a patrol returning to the JSS was confronted once again by an RKG-3 ambush, but this time things were different. With the experimental equipment and innovative tactics, the 5-4 Cavalry patrol defeated the RKG-3 ambush. Statistics tracked across all Iraq had documented 79 RKG-3 attacks between January 2008 and January 2009. USF win-loss record against the RKG-3 ambush to this point was 0-79. The USF recorded its first tactical victory against the RKG-3 that fateful night when a unit had adapted. The enemy was shot before he was able to throw the RKG-3. Severely wounded by an M4 carbine round in the gut and shotgun wounds to the arm, the enemy was saved by an American medic, and he was taken prisoner. A feeling of elation was felt across the organization, its sentiments captured by the squadron's executive officer that night in an email he sent out to the organization:

Tonight, I observed a patrol staged in JSS Ghaz-1 that had just returned from yet another RKG-3 attack in Ghazaliyah. There were very few similarities from this patrol and the first three attacked patrols that had returned to Ghaz-1 since our RIP/TOA [relief in place/ transfer of authority]. This time, there were no pockmarked or destroyed vehicles; there were no medics and litter bearers scrambling to treat the injured; there were no worried and apprehensive looks. Tonight's attack had profoundly different results because of our unit's ability to evolve against a new and emerging threat.[15]

Major David Dunphy, Squadron XO,
February 5, 2009

In contrast to the successful adaptations as described above, the last two successful RKG-3 attacks in Baghdad/Abu Ghraib were on January 29 and February 1, 2009. It is important to note the TTPs in use in the first video by a transit unit moving through 5-4 Cavalry's AO. The unit is not using new counter-RKG-3 TTPs: In the last vehicle driving at 25-30 mph (5-4 Cavalry reduced in-city driving speed to a hunting pace of 15-18 mph), the gunner surrounded by armor has no situational awareness — facing to the rear with his .50 caliber machine gun raised, using old Cold War air-guard tactics as if a Soviet Hind helicopter may sneak up on him in Iraq. Using these TTPs, the gunner and his crew have no idea what is about to hit them.[16]

CHANGING CULTURE THROUGH SUCCESSFUL ADAPTATION

Success bred success, and the unit embraced the experimental modifications and innovative tactics and aggressively trained them to perfection. The new TTPs became accepted as "how we do business around here." The squadron changed at this point. This adaptive challenge had been met with an effective and imaginative adaptive solution, and the culture of the entire organization changed. The culture of experimentation and innovation took root, and success of imagination began to spill over to all endeavors. Taking risk for the opportunity of a major payoff became the norm. It ignited a new offensive spirit that had been lost after years of base defense and counter-IED TTPs.

The tactics were developed to regain tactical initiative and achieve force protection. Stated another way, "the best defense is a good offense," with the original end state being to protect the patrols while moving from point A to point B, while conducting the squadron's primary mission of developing the ISF and improving civil capacity. The 5-4 Cavalry continued with this primary mission with new confidence that its patrols would win a tactical engagement with the RKG-3 ambush. Its new tactics became its trademark, and since the enemy knew when a 5-4 Cavalry patrol was passing, it experienced no further attacks. But the challenges were not over. The enemy adapted again: Don't attack the patrols with the modified turrets and new scanning techniques (5-4 Cavalry) . . . wait for a patrol from another unit passing through the AO who was using the old tactics.

The next adaptive challenge came on March 16, 2009, when a U.S. Army Military Police (MP) patrol was on the way from the forward operating base (FOB) at Camp Liberty en route to an Iraqi Police Station inside the 5-4 Cavalry AO to conduct Iraqi Police training and development. The patrol was hit by an RKG-3 ambush, killing one U.S. Army MP Soldier in the attack. Frustration again gripped the squadron. Despite its best efforts to share the gospel for how to maneuver inside their AO of the Mansour District and Abu Ghraib (RKG-3 "hot-spots") and attempts to proliferate the new TTPs used in their AO, many units who transited the 5-4 Cavalry AO were still using the TTPs for counter-IED operations and not counter-RKG-3 operations. This was by no means due to a lack of effort on the leadership's part.

The SCO/CSM "road show" brief that was used internally to educate the platoons in the squadron had been requested by the Commanding General of 1st Cavalry Division and posted on the division (MND-B) and corps (MNC-I) intranets for assimilation by all units in theater. The brigade and division commanders both made reference to the new tactics, and the expectation from them was that units that transited the 5-4 Cavalry AO (aka the RKG-3 threat areas) needed to read, learn, and implement these tactics. The squadron went on to publish a TTP article in the *Armor and Cavalry Journal* and communicated with the Combat Training Centers (CTC), which are responsible for training and certifying units coming to theater, explaining the newly developed TTPs for counter-RKG-3 for integration into future CTC rotations. Despite the best efforts of the chain of command, most units, because they had not truly dealt with the adaptive challenge, failed to change their behavior. The transfer of this implicit and explicit information met resistance and was often not employed by transit units in the 5-4 Cavalry AO.

The real frustration, though, was the fact that despite their ability to win the tactical engagement, they were not providing security for others transiting their AO. As the land owner or the maneuver unit responsible for the security and freedom of maneuver within its AO, 5-4 Cavalry took it personally that a U.S. Army Soldier was killed in their AO. This adaptive challenge was met with more innovation. The group's solution was to adapt again and take the new tactics on the offense. "Be the hunter, not the hunted" was no longer going to be a slogan for the "best defense. . . ." It would mean going on the offense. Search and Attack[17] operations commenced. When the 5-4 Cavalry patrols rolled out, **they gave the appearance of being a transit unit** by not overtly using counter-RKG-3 tactics, with the intent to entice the enemy to attempt an ambush so that the patrols could attack to destroy the enemy *before* transit forces moved into and through the AO. This allowed the patrols to maneuver unmolested. Providing security—a classic Cavalry mission!—5-4 Cavalry had the initiative and that spurred an offensive spirit.

The new offensive tactics were successful and kept the enemy on its heels by taking the initiative away from him and getting into *his* observe, orient, decide, and act (OODA) loop. Over the next 90 days, RKG-3 attacks declined by 90 percent, dropping from an average of 10 ambushes a month inside the 5-4 Cavalry AO to one attack a month by April 2009. The fact was, however, that the squadron had not destroyed the network responsible for the employment of the ambushes. That network could regenerate foot soldiers and conduct attacks in the AO against transit units despite being thoroughly disrupted by the 5-4 Cavalry TTPs. The next adaptive challenge was defeating the network behind the RKG-3. Innovative tactical patrol TTPs were not going to solve this challenge, but the experimental culture born during the development of these TTPs would be critical to overcoming the newest challenge of destroying the entire network.

Fighting networks was nothing new to USF and ISF forces by 2008-09. The Army and Marine Corps had met that adaptive challenge and over time the force made the necessary adjustments from defeating the regular forces of Saddam's Army to fighting irregular insurgent groups using networked cells to counter the strength of the U.S. military. **What the new culture of imagination and experimentation in 5-4 Cavalry allowed the squadron to do was to take this entrepreneurial spirit into its primary mission of partnership with the ISF.** The squadron was partnered with the 54th Brigade of the 6th Iraqi Army Division, and the two units teamed up to exploit the successful capture of the first defeated RKG-3 thrower and the subsequent intelligence derived from him and others over time as the network was attacked.

The 5-4 Cavalry and the 54th Iraqi Army Brigade **tore down traditional barriers** of mistrust and stereotypes typical between U.S. and Iraqi Forces, **and pushed the boundaries** of what and how intelligence could be shared among the combined forces. Prudent risk was taken in the sharing of intelligence, which resulted in a new level of trust, respect, and teamwork that ensued from the experiment. Combining the human intelligence capabilities of the 54th Iraqi Army Brigade and the technological intelligence capabilities of USF, the *1920's Revolutionary Brigade* in northwest Baghdad was ultimately destroyed.[18]

REFLECTION AND ANALYSIS: LEADERSHIP, RISK, AND PUTTING YOURSELF ON THE LINE

The exercise of leadership is a dangerous endeavor. The higher the stakes, the more challenging the task as the leader balances risk associated with lives, mission, and personal loss. Leadership requires moral courage and a willingness to take personal risk, to put oneself on the line. The defeat of the RKG-3 ambush is a success story because the experiment ended with a positive outcome. But leaders at all level accepted risk to allow for the experiment to take place. What if the outcome had been different, and the enemy had won and lives were lost? There were many factions waiting in the wings to discard those who had been willing to step across the line and take risks. In a hierarchical culture, leaders must be willing to take risks, allow for experimentation, and stimulate the creative process that generates discoveries.

This may place one in professional peril or even physical danger, but this is the essence of leadership. Soldiers must see you on the line both figuratively and literally. You may have to demonstrate personally the tactics and lead RKG-3 hunting expeditions, or you may have to underwrite and take responsibility for a subordinate's failure. In the case of breaking down barriers and pushing boundaries with the 54th Brigade, the commander role-modeled what was expected in this paradigm shift. In building the relationship with the commander of the 54th Brigade, the leader placed himself in vulnerable positions both personally and professionally by placing trust in our Iraqi ally. It went beyond sharing tea and exchanging pleasantries; it was putting his life and the lives of his men in the hands of the Iraqi commander, it was using discretion in what intelligence was shared, it was bending rules to build a relationship.

These were all risks. If they had gone badly, for example if a U.S. Soldier was killed while subjected to tactical decisions of the Iraqi commander or if classified information shared with the 54th Brigade was compromised, the U.S. commander assuming this risk could face reprimand or relief by factions less willing to be creative. In the end, the exercise of leadership is about setting conditions for progress. During times of adaptive challenges that require creative and imaginative problem-solving, it means taking risk and shaping the environment for the unit to conduct experiments. In a culture that is averse to failure, it requires the leader to accept personal risk and be prepared to assume the responsibility of failed efforts by subordinates using their imagination to find adaptive solutions. This willingness to assume risk must be tempered with an understanding of how to swim in the dangerous waters of leadership.

This point is so important that I will develop it by speaking directly to potential leaders in the second person: Leadership requires you to be present to exercise it. If you allow yourself to be marginalized or relieved, then you may have failed in your exercise of leadership. Be careful not to assume the role of the crusader or make yourself a martyr. Master the art of staying alive to fight another day. Pick your battles wisely, but when you do, be ready to step out on the line. There may be a time to fall on your sword, but understand you lose all say in the progress of the group if you are killed off and replaced by another who may not be willing to pick up the fight. One way to facilitate the wise acceptance of risk and to survive is to build alliances among factions in the system. Find partners and lead change with a united effort. This gives you more flexibility and increases the threshold of risk tolerance by those who find your prudent risk-taking uncomfortable or stressful.

The fact is that a network could be *defeated*[19] by disrupting its supply lines, or capturing elements of the cells, but could not be destroyed by USF efforts alone or limited combined operations under the old rules. This was because of the flat organizational structure and decentralized operating procedures employed by the Sunni insurgent cells and networks fighting in Iraq. Rarely is the network completely destroyed. In the present case there was no failure of imagination. With experimental combined (USF and ISF) operations the entire network was dismantled piece by piece, culminating with a combined effort to capture the enemy battalion commander of the *1920's Revolutionary Brigade*, a former major in Saddam's Army whose area of operation included northwest Baghdad (Mansour, Ghazaliyah, and Abu Ghraib) and who had overseen 53 RKG-3 ambushes between September 2008 and July 2009 (providing mission command and logistical support, he was not an executor/thrower).

This insurgent leader would later brag to ISF interrogators that his cell had gone 5 years without being defeated by Coalition Forces (he was unaware of the USF collaboration in his capture and thought the ISF alone had uncovered him). This insurgent leader boasted in his confession to the ISF that his *1920's Revolutionary* cell conducted the IED campaign on Route Irish in 2003-04, the most dangerous road in Iraq during that period, and touted the IED and sniper operations along Route Sword that claimed a number Coalition lives from 2005 to 2007. Cross-checking his interrogation results against historical attack data confirmed many of his claims. His capture by elements of the 54th Iraqi Army Brigade with USF support ended all RKG-3 ambushes across Multi-National Division-Baghdad from that point forward. Due to the horizontal organization of this particular insurgent group, the capture of its leader completely destroyed the *1920's Revolutionary Brigade*, ending its 5-year reign of anti-Coalition operations in northwest Baghdad.

At this juncture, it is important to acknowledge that 5-4 Cavalry was a system within a larger system and that the larger system played a significant role in the subsystem's ability to be adaptive. The command climate of the larger systems determined the level of discretion and flexibility the SCO and squadron were permitted to experiment and conduct adaptive work. Despite 5-4 Cavalry's immediate higher headquarters (2nd Brigade of the 1st Infantry Division) providing a command climate supportive of risk-taking, that was not the case at the beginning of the deployment with regard to the next higher level headquarters (Division).

For the first 3 months, the 2nd Brigade, 1st Infantry Division, and its subordinate units (to include 5-4 Cavalry) served under the command of 4th Infantry Division (4th ID). In addition to the SCO's own failure of imagination, the squadron was also immersed in this larger system, which was characterized by a hostile command climate in 4th ID—commanded by a division commander whose caustic style of leadership produced risk aversion and paralyzed creative thought. This period corresponds directly to the period when the subsystem had its failure of imagination and struggled with overcoming its adaptive challenge.

The 1st Cavalry Division (1st CD) transitioned into MND-B and took over for 4th ID. In contrast, under the 1st CD leadership of Major General Dan Bolger and Command Sergeant Major Rory Malloy, the climate was such that experimentation and initiative were not only tolerated, but encouraged and rewarded. The climate established within this larger system by the 1st CD command team provided the conditions for learning and adapting, thus significantly contributing to the ability of the subordinate organizations to display operational adaptability. When 5-4 Cavalry prepared to depart Baghdad at the end of its 12-month deployment, Major General Bolger invited the troop commanders and troop first sergeants of 5-4 Cavalry to the division headquarters for dinner to express his pride and admiration in their performance and initiative. His parting words to those junior leaders capture the essence of overcoming a failure of imagination to conduct adaptive work: "Yes Long Knives [5-4 Cavalry call sign], you *out-fought* the enemy here in Baghdad, but the real reason you won, and won decisively, is because you **out-thought** the enemy."[20]

LESSONS LEARNED FROM LEADING A LEARNING ORGANIZATION

The current and future operational environments that the U.S. Army can expect to face in this globalized and networked world will present ambiguous and complex problems that will not always allow commanders to reach into a bag of tricks and pull out a suitable solution. For a commander, there will be times when routine problems requiring a technical solution or the application of a standard operating procedure will indeed apply, and as the authority you will be able to apply your experience and expertise to solve the organization's problem. There will also be times, however, when your authority alone will be insufficient to produce the best course of action or the right solution to an adaptive challenge. In these times you, the commander, the platoon leader, or section sergeant must exercise adaptive leadership. Stated another way, you must mobilize the group to face the reality of a changed environment. You must be a coach, an interpreter, a guide in helping **the group** to conduct adaptive work.

This is not easy, as oftentimes it requires individuals and the group to challenge assumptions, abandon entrenched beliefs, and change cultural norms that are deeply rooted and have in the past proven to be successful. The more challenging the problematic reality, the more the group will want to cling to what worked in the past and resist change and uncertainty. The organization will want to avoid disequilibrium and will look to the leader to provide a solution that will reestablish the equilibrium quickly and painlessly. Unfortunately, easy and readily available answers are usually the wrong solution for a complex problem, especially if it requires adaptation by a group.

Often complex problems call for real leadership and require the person in a position of authority, like a commander, to avoid providing counterfeit solutions to appease the group and instead go against his/her tendency and natural desire to be a problem-solver. The authority may actually have to generate disequilibrium to mobilize the group to conduct adaptive work. To do this, the leader must first be reflective and be able to pull him/herself out of the fray to listen, sense, think, and then diagnose the problem. Then, if adaptive work is to be done, the leader must mobilize the group to take risk, experiment, innovate, and produce change — change that may be met with resistance and scorn from below and above. It is hard work to exercise leadership when confronted with adaptive challenges. It takes real leadership.

The exercise of leadership goes beyond the mere possession of authority. Unfortunately, all too often commanders rely only on authority itself, including reliance on a repertoire of known solutions. This is a risk-averse approach to command, an approach that avoids creating disequilibrium. But it is in states of disequilibrium where the most effective social learning takes place. With this risk-averse approach, at the end of your 12-month tour you will be able to say to yourself, "We did what we trained to do. Yes, we took casualties, but not because we were deficient in our execution of approved solutions; the enemy just got the best of us that day. . . . We didn't destroy the enemy, but we feel that we disrupted him thoroughly. . . . So I know in my heart we did our best." Did you? Or did you just lack imagination? Did you fail to adapt or adapt enough? Did you take on the challenge of leading your organization to face the reality that the environment had changed and what you trained to perfection for 12-months before your deployment is less relevant now that you are in the arena of *today*, where in-stride changes and adaptations have to be made? Did you have the moral courage, while in contact with the enemy, to create disequilibrium in your organization so it could do real adaptive work? Did you create a culture in which your organization became a learning organization? Or did you lack imagination, fail to revisit assumptions, engage in work avoidance to sidestep the real work, and provide counterfeit leadership from a position of authority?

These are the hard questions that commanders need to ask themselves as they continue to face complex and ambiguous operational environments. Despite the superior comparative strength in military capabilities of the U.S. Army, it is the failure of imagination that can be the U.S. Army's Achilles heel. The Army's competitive advantage must come from its intellectual approach to future conflict. It will not only be the Army's ability to out-fight the enemy, but to out-think him, to adapt faster, and to seize and maintain the initiative, that will produce future victories.

ENDNOTES - CASE STUDY

1. *National Defense Strategy 2008*, Washington, DC, p. 19, available from *www.defense.gov/news/2008%20national%20defense%20strategy.pdf*.

2. **Counterfeit leadership**, as defined by Dean Williams, is the kind of action, irrespective of one's intentions, that results in putting a false set of tasks before people, including any activity pursued by a group that has nothing to do with progress: false strategy, political game playing, interdivisional rivalries, tolerance of scapegoating, and refusal to admit error and learn.

3. **Adaptive work**, as defined by Dean Williams, is the effort that produces the organizational or systemic learning needed to tackle tough problems that often require an evolution of values, beliefs, or practices.

4. Information on this weapon system and the enemy adaptive tactics is available from *www.liveleak.com/view?i=df5_1226223621*.

5. **Authority**, as defined by Dean Williams, is the power granted to perform a service. Formal authority arises when the officeholder promises to meet a set of explicit expectations, such as a job description or professional standards. This exchange of power for a service may be informal. Informal authority arises when employees confer power on a person based on implicit expectations.

6. **Technical Solutions**, as defined by Dean Williams, is an intervention plan that includes: (1) a fixed statement of the problem, and, (2) fixed standards for success defined before the intervention was made. Routine solutions for routine problems. Solutions not requiring learning or adapting.

7. John B. Richardson IV, *Command Guidance*, Baghdad, Iraq: Headquarters 5-4 Cavalry, January 2009.

8. The TTPs developed and implemented can be found in an article published in the May-June 2009 edition of the *Armor and Cavalry Journal,* or through the Center for Army Lessons Learned.

9. U.S. Army, 5th Squadron, 4th Cavalry Command Brief, "Defeating the RKG-3," January 5, 2009.

10. *Ibid.*

11. The OODA loop is a concept originally applied to the combat-operations process, often at the strategic level. It is now also often applied to understanding commercial operations and the learning processes. The concept was developed by military strategist Colonel John Boyd, USAF.

12. **Movement to Contact** is a form of the offense designed to develop the situation and to establish or regain contact with the enemy.

13. **Meeting Engagement** is a combat action that occurs when a moving force, incompletely deployed for battle, engages an enemy at an unexpected time and place.

14. **Ambush** is a surprise attack by fire from concealed positions on a moving or temporarily halted enemy.

15. John B. Richardson IV, "Be the Hunter, not the Hunted: Adapting to Defeat the RKG-3,"*Armor and Cavalry Journal*, May-June 2009, p. 5.

16. These videos are available from *www.archive.org/details/1920RevolutionBrigades-NewRkg-3GrenadeOnUsHumvee*, and *www.archive.org/details/1920RevolutionBrigades-Rkg-3GrenadeDamageOneStryker*.

17. **Search and Attack** is a variant of the movement to contact conducted by smaller, light maneuver units and air cavalry or air assault forces in large areas to destroy enemy forces, deny area to the enemy, or collect information. Search-and-attack operations may be conducted against a dispersed enemy in close terrain unsuitable for armored forces, in rear areas against enemy special operations forces (SOF) or infiltrators, or as an area security mission to clear assigned zones.

18. **Destroy** is defined as: (1) a tactical task to physically render an enemy force combat-ineffective unless it is reconstituted,

or, (2) to render a target so damaged that it cannot function as intended nor be restored to a usable condition without being entirely rebuilt.

19. **Defeat** is a tactical task to either disrupt or nullify the enemy force commander's plan and subdue his will to fight so that he is unwilling or unable to further pursue his adopted course of action and yields to the will of his opponent.

20. Major General Daniel Bolger, speech to commanders and first sergeants of 5th Squadron, 4th Cavalry, Baghdad, Iraq, September 2009.

CHAPTER 1

THE ARMY'S CAMPAIGN OF LEARNING

> Leaders lacking adaptability enter all situations in the same manner and often expect their experience in one job to carry them to the next. Consequently, they may use ill-fitting or outdated strategies. Failure to adapt may result in poor performance in the new environment or outright organizational failure.[1]
>
> Field Manual 6-22, *Army Leadership*

As the U.S. Army prepares for the second decade in the era of persistent conflict, it is incumbent on the profession of arms to reflect on the previous decade of war and learn the appropriate lessons. This reflection allows leaders in the Army to diagnose, challenge, and intervene in the system to promote change in order to revamp the organizational culture and maximize the growth in its next generation of military leaders, and to strengthen their profession as it seeks to improve its mastery and management of lethal force.[2] In 2009, TRADOC launched an introspective analysis, whose thesis is the "campaign of learning—a set of initiatives built on the expectation of persistent conflict, grounded in the lessons learned from 9 years of war, and balanced against the emerging trends of the future operational environment."[3]

THE OPERATIONAL ENVIRONMENT

We have always lived in uncertain, unpredictable, and challenging times, but as a result of globalization and proliferation of technology, today's uncertainty is fundamentally different. The security environment is much more competitive in the post-September 11, 2001 (9/11) world. Countering the Soviet conventional threat was, in theory, exploiting the predictability by which the Soviet armored columns would attack. While serving as a tank platoon leader in Germany during the early 1990s, my ability to display conceptual agility was not an imperative, just my ability to do math. It was "battlefield calculus" — there are this many of them, coming in this manner, at this rate, and my platoon can kill this many, at this distance, at this rate. The paradigm was simple, with solutions generated from a repertoire of time-proven battle drills and standard operating procedures.

Today's tank platoon leader lives in a different world. As General Martin Dempsey explains, it is a world where "uncertainty is the result of persistent conflict with hybrid threats, enabled by technology, that decentralize, network, and syndicate . . . an environment where we should expect to be surprised more frequently and with potentially greater impact."[4] As outlined in the *Army Capstone Concept*, "The ability to adapt depends on a fundamentally sound estimate of future threats, challenges, and enemy capabilities as well as an understanding of the future operational environment."[5] In simple terms, the future operational environment (OE) will exhibit uncertainty and complexity. The future OE will be highlighted by asymmetric tactics employed by "hybrid enemies: both hostile states and non-state enemies that combine a broad range of weapons capabilities and regular, irregular, and terrorist tactics; and continuously adapt to avoid U.S. strengths and attack what they perceive as weaknesses."[6]

The OE requires leaders at all levels to work within the context of a host-nation/multinational construct.

The OE will be focused on populations, thus presenting challenges in lingual and cultural differences. The enemy will be networked and not always rational. The host nation and the enemy will be news-media savvy and will use information operations to achieve their ends. Increasingly, in this networked and globalized world, the threat will operate in flat organizations. As Ori Brafman and Rod Beckstrom point out in their book, *The Starfish and the Spider: The Unstoppable Power of Leaderless Organizations*, to combat the strengths of the United States, the threats in the networked world will have minimal hierarchy, making their organizations highly adaptive and decentralized for rapid decisionmaking.[7] Recent examples of this include Sunni insurgent groups in Iraq, Hezbollah in Lebanon, and al-Qaeda (AQ) around the globe. The bottom line is that the OE will remain ambiguous and complex, with a highly adaptive enemy. It is this ever-changing environment that requires a fundamental shift in the approach to military problem-solving and Army leader development.

ARMY INITIATIVES

Seismic shifts in how the Army operates after a decade of combat against irregular forces and networked cells of insurgents employing asymmetric tactics demonstrate that **the Army can adapt**. Like General George Casey, Chief of Staff of the Army from 2007 to 2011 suggests, the real question is, how quickly can we adapt and at what cost?

General Casey concluded his annual assessment in the 2009 *Army Green Book* as follows:

The Army of the 21st century described here will require continuous change. Our strategic environment has evolved dramatically, and so has the Army. The challenges of institutional change in large organizations like the Army are substantial, especially as we are adapting an organization that is already the best in the world at what it does. Our test must not be, "Have we changed?" It must be, *"Have we changed enough?"* Everything is on the table except our core values.[8]

Senior Army leaders' reflections on decade of war illuminated the need for the Army to codify new practices, values, and procedures. The U.S. Army Training and Doctrine Command (TRADOC) took the lead in redesigning the Army's conceptual foundation by reflecting on the previous decade of persistent conflict and is aggressively pursuing a *campaign of learning*, "a broad set of initiatives designed to produce an Army capable of rapidly adapting to defeat unforeseen threats."[9] The Army's Campaign of Learning is General Dempsey's *strategic intervention* to promote change.

The common thread in all the following developments and other TRADOC initiatives is the requirement for continuous learning and adaptability. It is a recognition that the ability to frame complex problems, to understand the problem in context, and to adapt quicker than the enemy can, is what will give the U.S. Army a competitive advantage on the battlefields of the 21st century. In essence, it is the ability to learn operational adaptability.

A brief description of some of TRADOC's most significant contributions to the learning process reveals that the Army — as a learning organization — is making remarkable strides within the problem-solving process. The Army Capstone Concept (ACC) serves as the intellectual foundation of the Army concept framework, setting forth the Army's strategic vision of future armed conflict. The central theme of the ACC is operational adaptability, considering future armed conflict within the context of four key trends in the current and future operating environment: uncertainty, pace of change, competitiveness, and decentralization. The ACC also identifies the substantive adaptations that we must make in how we build leaders, and how we train, learn, and organize ourselves.

The Army Operating Concept (AOC) describes "how" future Army forces conduct operations as part of the Joint Forces to deter conflict, prevail in war, and succeed in a wide range of contingencies, all as focused on the employment of forces in time frame 2016-28. The ideas brought forth in the AOC will guide future revisions in Army doctrine, organization, training, materiel, leader development, education, personnel, and facilities. The AOC also elucidates the necessity of operational adaptability in full-spectrum operations against hybrid threats with the land warfare capabilities to conduct combined arms maneuver and wide area security.

The Army Leader Development Strategy (ALDS) seeks to establish the right balance between the three pillars of leader development: training, education, and experience. The ALDS describes three required paradigm shifts as the Army adapts its leader development model: 1) the effect of complexity and time on decisionmaking; 2) the effect of decentralization on organizational leadership; and, 3) the need for the ability to frame ill-structured problems. An excerpt from ALDS highlights the following attributes and core competencies sought in U.S. Army leaders:

LEADER ATTRIBUTES

Army leaders must possess and model key attributes in order to reach their full professional potential. An attribute is defined as a characteristic unique to an individual that moderates how well learning and performance occur. Leader development must build on the foundation of an individual's existing qualities, developing well-rounded leaders who possess three critical leadership attributes.

Character. A leader of character internalizes the Army Values,[1] lives by our Professional Military Ethic, reflects the Warrior Ethos,[2] and displays empathy toward Soldiers, families and those people affected by the unit's actions. Character is central to a leader's core identity. In our profession, competence places an individual in position to lead — character makes him or her an effective leader.

Presence. A leader of presence has credibility, exudes confidence, and builds trust. Presence is conveyed through actions, appearance, demeanor, and words.

Intellect. A leader of intellect has the conceptual capability to understand complex situations, determine what needs to be done, and interact with others to get it done. Leaders must have the ability to reason, to think critically and creatively, to anticipate consequences, and to solve problems.

---

1 **Army Values**: Loyalty, Duty, Respect, Selfless Service, Honor, Integrity, and Personal Courage.

2 **Warrior Ethos**: I will always place the mission first; I will never accept defeat; I will never quit; I will never leave a fallen comrade.

LEADER CORE COMPETENCIES

Army leaders apply their character, presence, and intellect in leading our nation's Soldiers. The expectations for what leaders should do regardless of the situation are captured in the Army's core leader competencies. Core leader competencies are defined as groups of related behaviors that lead to successful performance, common throughout the organization and consistent with the organization's values:

One who leads. Provides vision through purpose, motivation, universal respect, and direction to guide others. Extends one's influence beyond the chain of command to build partnerships and alliances to accomplish complex work. Leading is conveyed by communicating (imparting ideas) and setting the example.

One who develops. Leads organizations by creating and maintaining a positive environment and by investing effort in their own broadening and in that of others to achieve depth and breadth. Developing includes assessing needs to improve self, others, and the organization.

One who achieves. Focuses on what needs to be accomplished. Has an expeditionary mindset and can adapt to unanticipated, changing, and uncertain situations. Achieving in the short term is about getting results, but in the long term it is about setting the vision to obtain objectives.[10]

Additionally TRADOC is spearheading the Army Learning Concept for 2015 (ALC), designed to win in the competitive learning environment by creating an

49

accessible, responsive, career-long continuum of learning that supports operational adaptability. Additionally, the Army Training Concept 2015 (ATC) provides a vision for the Army's way ahead for unit training, which strikes a balance between operational and institutional training requirements and offers flexibility, efficiencies, and a broad range of training capabilities so leaders can maintain operational adaptability.[11]

Finally, TRADOC is refashioning key doctrine to include *Field Manual (FM) 3.0, Operations,* which codifies the hybrid threat as critical to the commander's understanding of the complex combination of threats the Army faces. FM 3.0 also integrates the doctrinal concept of mission command—the adaptation that captures what we have learned in 9 years of war, including roles and responsibilities of leaders in distributed and increasingly decentralized operations. The new *Field Manual (FM) 7.0, Training Units and Developing Leaders for Full Spectrum Operations,* supports FM 3.0 and encourages an intellectual rather than a lock-step management process by providing broad unit training and leader development concepts and encouraging the use of mission command in training.

OPERATIONAL ADAPTABILITY

Operational Adaptability emerges from these concepts and is the cornerstone of successful application and integration of the military component of national power at the tactical, operational, and strategic levels of war. It is a quality based on critical thinking, comfort with ambiguity and decentralization, a willingness to accept prudent risk, and an ability to make rapid adjustments based on a continuous assessment of the situation.[12]

Application of this concept assumes that the Army culture is conducive to innovation and experimentation, and rewards prudent risk takers, and that based on this culture, those in positions of authority in the Army are adaptive leaders with the skills and attributes to lead organizations through problematic challenges[13] requiring adaptive work.[14]

As the Army attempts to move toward embracing operational adaptability, the question remains, has it as a profession made the necessary changes to its culture to embrace fully the concepts espoused by the emerging doctrine? Does the organizational culture nurture the growth of this type of leader? No one in today's Army will deny the need to achieve operational adaptability, but aligning those espoused values with basic underlying assumptions is an adaptive challenge the Army faces in fully incorporating new behavior into its culture.

Acknowledgment of the changed OE and of the need to achieve operational adaptability to survive and win leads to the question: What type of leader does the Army need to perform effectively in the context of operational adaptability? General Casey answers these questions when he stated: "Our Soldiers must be led by agile, culturally astute, and adaptive leaders."[15] Based on the current OE, our profession demands leaders with greater imagination and increased awareness of "weak signals" of impending change.[16] According to General Dempsey, "confronting hybrid threats . . . in such an environment requires leaders who not only accept but seek and embrace adaptability as an imperative."[17] In response to the emerging trends in the OE, he goes on to state, "Our profession must embrace a culture of change and adaptation. We must think differently about how we develop leaders."[18]

ENDNOTES - CHAPTER 1

1. *Field Manual (FM) 6-22, Army Leadership*, Washington, DC: HQ Department of the Army, 2006, pp. 10-18.

2. John B. Richardson IV, "Shaping Adaptive Leaders for the 2nd Decade and Beyond" (pending publication), p. 1.

3. Martin E. Dempsey, "Driving Change Through a Campaign of Learning," *Army Magazine*, Vol. 60, No. 10, October 2010, p. 66.

4. *Ibid.*

5. TRADOC *Pamphlet (PAM) 525-3-0, The Army Capstone Concept*, Washington, DC: HQ Department of the Army, 2009, p. 15.

6. *Ibid.*

7. Ori Brafman and Rod A. Beckstrom, *The Starfish and the Spider: The Unstoppable Power of Leaderless Organizations*, New York: Penguin Group, 2006, p. 134.

8. George W. Casey, *Army Magazine*, Vol. 59, No. 10, October 2009, p. 40.

9. TRADOC Themes and Messages, February 3, 2010.

10. Army Leader Development Strategy, November 2009, p. 9, available from *www.nps.gov/abli/*.

11. TRADOC Themes and Messages.

12. PAM 525-3-0, p. 16.

13. **Problematic challenge,** or adaptive challenge, is a problem that resists solution even when leaders apply the best-known methods and procedures based on the current values in the organization; generally, the resolution of an adaptive challenge requires a shift in values.

14. PAM 525-3-0, p. 209.

15. Casey, p. 28.

16. Dempsey, "Driving Change Through a Campaign of Learning," p. 66.

17. Martin E. Dempsey, "Mission Command," *Army Magazine*, Vol. 61, No. 1, January 2011, p. 44.

18. Martin E. Dempsey, "A Campaign of Learning to Achieve Institutional Adaptation," *Army Magazine*, Vol. 60, No. 11, November 2010, p. 34.

CHAPTER 2

THE ARMY'S ADAPTIVE CHALLENGE

Creativity provokes the jealousy of the gods.[1]

Rollo May

In his speech to the cadets at West Point, Secretary of Defense Robert Gates stated that the Army faces the challenge of adapting its practices and culture to the strategic realities of the 21st century. "How," he asked, "could it better prepare itself, and in particular its leaders, for a complex and uncertain future?"[2] He continued, "From the beginning of the wars in Iraq and Afghanistan, our Soldiers and junior- and mid-level leaders down range have been adjusting and improvising to the complex and evolving challenges on the ground."[3] But he emphasized that it had taken the bureaucracies here at home longer to "respond with remotely similar agility."[4]

William Deresiewicz, a former professor at Yale, explained the "great mystery about bureaucracies" in an earlier lecture to the cadets as he challenged them to be "thinkers" and to avoid the tendency to be consumed by the bureaucracy they are about to join. In reference to American leadership, Deresiewicz stated:

> For too long we have been training leaders who only know how to keep the routine going. Who can answer questions, but don't know how to ask them. . . . Who think about how to get things done, but not whether they're worth doing in the first place. What we have now are the greatest technocrats in the world. He asks: [W]hy is it so often that the best people [in a bureaucracy] are stuck in the middle and the people who are

running things—the leaders—are the mediocrities? Because excellence isn't usually what gets you up the greasy pole. What gets you up is a talent for maneuvering. Kissing up to the people above you, kicking down the people below you. Pleasing your teachers, pleasing your superiors, picking a powerful mentor and riding his coattails until it's time to stab him in the back. Jumping through hoops. Getting along by going along. Being whatever other people want you to be, so that it finally comes to seem that . . . you have nothing inside you at all. Not taking risks, like trying to change how things are done or question why they're done. Just keep the routine going.[5]

Such is the dire state of leadership in a bureaucracy.

Colonel Scott Krawczyk, a professor at West Point, here describes the model for officers as espoused early in our country's history:

> From the very earliest days of this country, the model for our officers, which was built on the model of the citizenry and reflective of democratic ideals, was to be different. They were to be possessed of a democratic spirit marked by independent judgment, the freedom to measure action and to express disagreement, and the crucial responsibility never to tolerate tyranny.[6]

This model contrasts markedly with the career pattern of the self-serving bureaucrat in uniform, General Courtney Massengale, as portrayed in Ken Follett's fine World War II novel, *Once an Eagle*. Massengale's foil is General Sam Damon, who spurns careerism and instead insists on being one of those officers who "think for themselves and act on their convictions."[7]

As Brigadier General H. R. McMaster once wrote, "Commanders and senior civilian officers must be willing to underwrite mistakes, mistakes of commission should be tolerated, passivity should not."[8] Lieutenant General George S. Patton, Jr., once wrote: "Collins and Bradley are too prone to cut off heads. This will make division commanders lose their confidence. A man should not be damned for an initial failure with a new division. Had I done this with Eddy of the 9th Division in Africa, the army would have lost a potential corps commander."[9] It is the fear of having your head cut off for making a mistake or challenging the status quo that hampers the imagination of leaders and generates a culture of risk aversion.

During recent visits to Harvard's Kennedy School, senior military leaders have voiced concern over the challenge of growing adaptive leaders. Lieutenant General Robert Caslen, commander of the Combined Arms Center at Fort Leavenworth, KS, stated that "his primary concern is the culture of risk aversion that could pervade the force."[10] This potential aversion to taking prudent risk is exacerbated by the way junior leaders perceive how risk-taking is promoted in the current operating environment (OE), while the corresponding responsibility and accountability are not being managed to allow for mistakes. This perception is reinforced by the handling of incidents like Wanat and COP Keating in Afghanistan. At these remote Army outposts, relatively high U.S. casualties resulted from clashes with Taliban forces. The ensuing investigations into the incidents resulted in reprimands for junior and mid-level leaders, which was an example, as perceived by many of those less-senior leaders, of senior Army leaders taking punitive actions against junior leaders for taking risks and making decisions. These perceived signals could result in a paralysis that runs counter to the espoused values of emerging doctrine.

There are examples of zero-defect command climates in our Army's past that provide lessons for the Army's leadership in managing risk in the 21st century. Dan Bolger in his 1990 Command and General Staff College paper entitled, "Zero Defects: Command Climate in First Army, 1944-1945," cites historian Russell Weigley in highlighting that the "underlying tactical weakness that precipitated the major crisis in the First Army" was based on "'unimaginative caution' as the overriding trait of these U.S. commanders. Most First Army generals showed themselves 'competent but addicted to playing it safe.' By comparison, Patton's Third Army and Simpson's Ninth risked more and accomplished more with significantly fewer losses."[11]

Brigadier General McMaster, former director of the Army Capabilities Integration Center's (ARCIC) Concepts Development and Experimentation Directorate, who is also the principal author of the Army Capstone Concept, stated,

> We need to reject the assertion that future war will differ fundamentally from recent and ongoing conflicts in order to protect future commanders from what could become a tendency toward risk aversion and over-control. Assuming information superiority might lead some commanders to conclude that making near-perfect decisions based on near-perfect intelligence is the essence of command. Commanders must be capable of conceptual thought and have the ability to communicate a vision of how the force will achieve its objectives.[12]

When asked what the Army's biggest challenges are as it prepares to tackle the future, General Ray Odierno, present Army Chief of Staff and former commander of Coalition Forces in Iraq, stated that "leaders will make the difference. We can build and field equipment quickly if we have to, but what we can't build quickly are leaders . . . and we need leaders who think differently and are creative."[13] Overcoming a risk-averse culture and imbuing the attribute of creativity in leaders is an adaptive challenge for the Army. It is an adaptation that will require significant disequilibrium within the institution because it will involve questioning taken-for-granted beliefs and ultimately changing organizational culture.

In his study of the Army's organizational culture, Dr. James Pierce "examines the degree of congruence between the Army's organizational culture and the leadership and managerial skills of its officer corps senior leaders." At the macro level, the results of his research strongly suggest a "significant lack of congruence between the U.S. Army's organizational culture and the results of its professional development programs for its future strategic leaders."[14] The study suggests that what the Army says is important is not believed to be truly important by the leaders in the Army.

ORGANIZATIONAL CULTURE

In general, organizational culture is considered to be very stable and difficult to change because it represents the collective repertoire of thinking, feeling, and perceiving that has enabled the organization to adapt successfully in reacting to internal and external environmental stimuli over a long period.[15] Edgar Schein, in his book *Organizational Culture and Leadership*, describes three levels of culture: **artifacts** that are visible symbols of a culture (uniforms, saluting, language, etc.); **espoused values**, which provide organizational members a sense of what ought to be; and finally, **basic underlying assumptions,** which evolve from a continuous use of a problem solution that has repeatedly been successful in the past and has unconsciously become taken-for-granted as the only way to solve similar problems. Therefore, organizational members instinctively perceive these basic underlying assumptions as "non-confrontable and non-debatable."[16] In essence they define "how we do business around here."

Schein emphasizes that the essence of an organization's culture is its taken-for-granted underlying assumptions, which provide consistency, order, structure, boundaries and ground rules, membership criteria, communication patterns, conditions for rewards, punishment, and the use of power.[17] Like Dr. Pierce's iceberg metaphor (See Figure 2-1), the true depth and breadth of an organization's culture lie beneath the surface and are very difficult to perceive through superficial analysis.[18] As a result of its hidden and often subconscious nature, an organization's culture makes bringing to light the realities of incongruence between espoused values and basic underlying assumptions a most difficult adaptive challenge. Organizational culture is a critical factor in the long-term effectiveness and survival of organizations. Consequently, senior leaders who provide strategic direction and vision for their organization must not underestimate the importance of culture and must realize that they are responsible for the analysis and management of their organization's culture.[19]

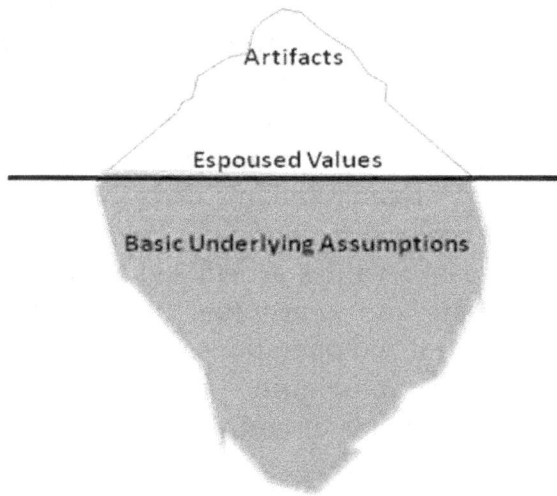

Source: James G. Pierce, *Is the Organizational Culture of the U.S. Army Congruent with the Professional Development of its Senior Level Officer Corps?* Carlisle, PA: Strategic Studies Institute, U.S. Army War College, September 2010, p. 11.[20]

Figure 2-1. Iceberg Conceptual Model of Schein's Three Levels of the Manifestation of Organizational Culture

THE REALITY FACING THE ARMY

Some have asked why we have placed increased emphasis on adaptation. . . . Several trends have emerged over the past decade . . . these trends have created an OE [operational environment] that is very dangerous, increasingly competitive, and always unpredictable. In response, our profession must embrace a culture of change and adaptation. We must think differently about how we develop leaders and how we organize, train, and equip our Soldiers and units.[21]

General Dempsey
November 2010

General Dempsey recently stated: "Aligning and connecting our leader development programs and policies with our conceptual foundation and doctrinal changes such as mission command become **the most critical adaptation we can make within our campaign of learning**."[22] Dean Williams describes this as "the work in the center."[23]

The espoused values articulated in the emerging Army doctrine and conceptual framework—adaptability, decentralized operations, discretion, initiative, and comfort with ambiguity—seem to be opposed to the basic underlying assumptions of Army leaders who, as Pierce explains, "emphasize stability, control, formalized structures, and a results-oriented— get the job done—culture that frequently attempts to comprehend the ambiguity of the future through an unconscious reliance upon the successful solutions employed in the past."[24] Surveys completed as part of Pierce's study by senior lieutenant colonels and colonels across the breadth of the force demonstrate this apparent lack of congruency. They paint a picture of a culture that values control over discretion and playing it safe over prudent risk-taking, despite what is espoused during formal and informal professional development discourse within the institutional Army. Pierce's study reveals that the officers surveyed believe that they operate on a day-to-day basis in an organization whose culture is characterized by:

- an overarching desire for stability and control,
- formal rules and policies,
- coordination and efficiency,
- goal and results oriented, and
- hard-driving competitiveness.[25]

However, sharply indicating a pronounced lack of congruence between what they believe the Army's culture to be and what it should be, the respondents also indicated that the Army's culture *should be* that of a profession, which emphasizes:

- flexibility and discretion,
- participation,
- human resource development,
- innovation and creativity,
- risk taking,
- long-term emphasis on professional growth, and
- the acquisition of new professional knowledge and skills.[26]

Recent documents such as the Army Capstone Concept (ACC), the Army Operating Concept (AOC), the Army Leader Development Strategy (ALDS), and *Field Manual (FM)-3.0, Operations,* espouse the need for leaders with skills and attributes found on the second list set forth above to combat the threats and complexity of the current OE. Culture, however, is not determined by what the profession says it wants its beliefs and values to be, but what it does, and that takes place in units on a day-to-day basis where informal professional development occurs. The incongruence in the Army culture highlighted by Pierce's research creates a "trust deficit" that militates against producing leaders with an entrepreneurial spirit. This disjunction is the heart of the adaptive challenge confronting the Army.

Dr. Pierce's analysis uses the Competing Values Framework (CFV) Model to evaluate the Army's culture because of its high degree of validity and reliability.[27] The CFV framework consists of four quadrants representing the four major cultural types—**clan**, **adhocracy**, **market**, and **hierarchy**—including an explanation of the differing orientations and competing values that characterize human behavior.[28]

In conjunction with the CFV, the Origins of the Organizational Culture Assessment Instrument (OCAI) allows for the diagnosis of the dominant orientation of an organization based on these core cultural types, cultural strength, and cultural congruence.[29] The results from Pierce's 2003-04 research indicated a lack of congruence that may be inhibiting performance and unconsciously perpetuating a cycle of caution and over-reliance on stability and control.[30]

To operate effectively in the current OE, companies, battalions, brigades, and divisions should strive to foster a culture that blends the characteristics of a "clan culture" and an "adhocracy culture." This blend moves toward flexibility and discretion (see Figure 2-2), thus ushering in a beneficial evolution in Army leadership.

Clan cultures are organizations held together by loyalty or tradition in which commitment to the organization is high. The organization emphasizes the long-term benefit of human resources development and attaches great importance to cohesion and morale. A clan culture places a premium on teamwork, participation, and consensus.

Adhocracy cultures include a dynamic, entrepreneurial, and creative place to work. People stick their necks out and take risks. The leaders are considered innovators and risk-takers. The glue that holds the organization together is commitment to experimentation and innovation. The organization encourages individual initiative and freedom.

Flexibility and Discretion

Culture Type: **CLAN**	Culture Type: **ADHOCRACY**
Leader Type: Facilitator, Mentor, Parent Effectiveness Criteria: Cohesion, Morale, Human Resource Development Management Theory: Participation fosters commitment	Leader Type: Innovator, Entrepreneur, Visionary Effectiveness Criteria: Cutting-edge output, Creativity, Growth Management Theory: Innovativeness fosters new resources
Culture Type: **HIERARCHY**	Culture Type: **MARKET**
Leader Type: Coordinator, Monitor, Organizer Effectiveness Criteria: Efficiency, Timeliness, Smooth functioning Management Theory: Control fosters efficiency	Leader Type: Hard-driver, Competitor, Producer Effectiveness Criteria: Market share, Goal achievement, Beating competitors Management Theory: Competition fosters productivity

Internal Focus and Integration (left axis) — *External Focus and Differentiation* (right axis)

Stability and Control

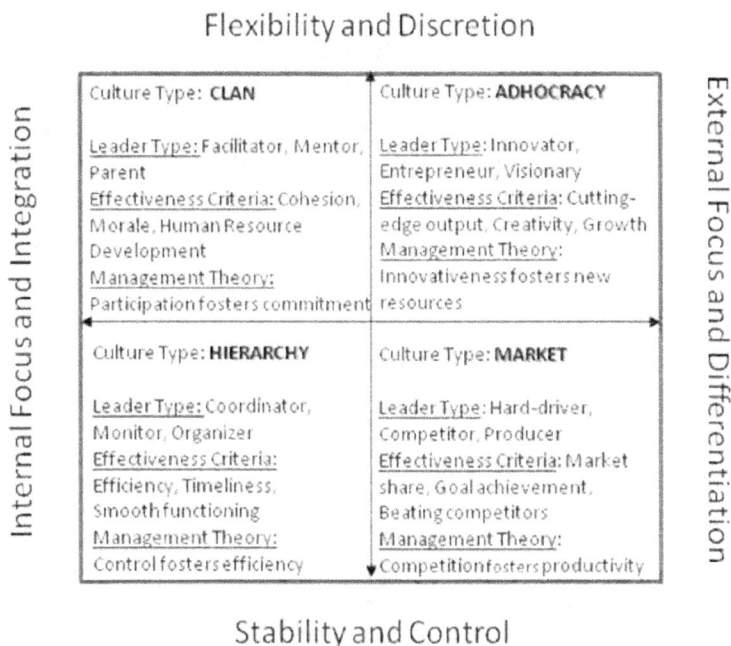

Source: James G. Pierce, "Is the Organizational Culture of the U.S. Army Congruent with the Professional Development of its Senior Level Officer Corps?" p. 53. [31]

Figure 2-2: A Summary of the Competing Value Set and Effectiveness Models.

Instead of cultures that epitomize flexibility and discretion, the bureaucratic nature of the Army can lead to the creation of a culture that seeks stability and control more commonly associated with **hierarchy cultures** and **market cultures** (see Figure 2-2). This is the challenge for the military professional, achieving adhocracy effectiveness and clan-like cohesion in a bureaucratic organization whose hierarchical chain of command is more conducive to a controlling culture of efficiency and risk aversion.

Figure 2-3 compares what the officers surveyed in 2003-04 believed the culture of the Army to be (represented as the "Now" U.S. Army culture), with what they believed the profession's culture should be (represented by the "Preferred" U.S. Army culture). These data highlight the gap between espoused values and basic underlying assumptions, and the challenge for Army leaders in reshaping the organization's culture.

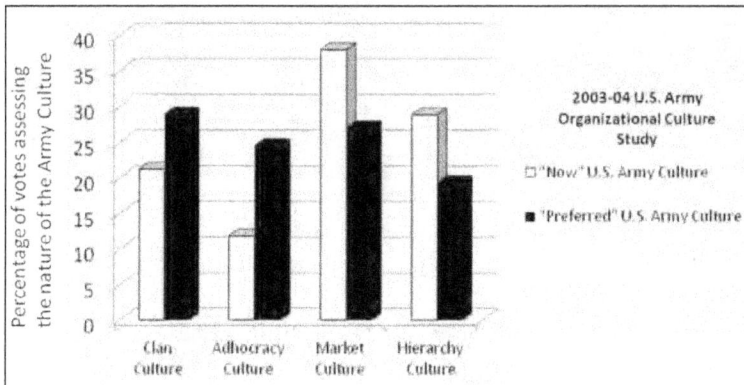

Source: James G. Pierce, data collection 2003-04, "Is the Organizational Culture of the U.S. Army Congruent with the Professional Development of its Senior Level Officer Corps?"[32]

Figure 2-3. "Now" vs. "Preferred" U.S. Army Culture. How the Officers Surveyed Perceive the Army's Culture *Is* ("Now") Compared to How they Think It *Should Be* ("Preferred").

Professor Douglas Lovelace, Director of the Strategic Studies Institute of the U.S. Army War College, summarizes Dr. Pierce's 2003-04 research:

Dr. Pierce's research data provide empirical support to the findings of the Army Training and Leader Development Panel (2001), which suggests that the training and *leader development programs of the Army profession are not adequately linked and integrated within the Army culture.* Dr. Pierce states that the Army's future strategic environment will be ambiguous and uncertain, and organizational culture and professionalism characterized by flexibility, discretion, and innovation offer the greatest opportunity to maximize effectiveness in such an environment. He postulates that if the Army profession expects to maintain its social legitimacy and professional jurisdiction, which are focused on the development and application of the esoteric knowledge and related practical professional skills of land warfare, then the Army profession must take steps to make its professional culture, and particularly the informal development program, congruent with one that is characterized by flexibility, discretion, and innovation.[33] (italics added by author)

After a decade of conflict in which adaptive behavior has been successfully employed on the battlefield, and 7 years since the previous research was completed, Dr. Pierce initiated a follow-up survey to see if the Army's culture had moved toward one more conducive to discretion and flexibility. His preliminary 2010-11 findings concerning the "Now" U.S. Army Culture figures are discouraging when compared to the 2003-04 study, as indicated below (See Figure 2-4).

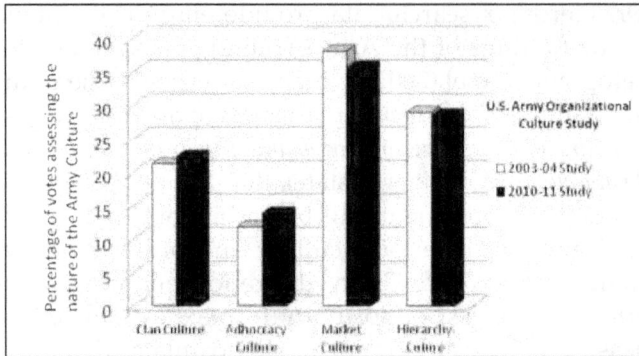

Source: James G. Pierce, data collection, comparison of Army Organizational Culture Data 2003-04 and 2010-11.[34]

Figure 2-4. "Now" U.S. Army Culture
How the Officers Surveyed Perceive the Army's Culture, Comparing Survey Data from 2003-04 and 2010-11.

What these figures show is that 7 years after the first study was completed, as perceived by 494 officers in the second survey, and despite a decade of persistent conflict, there has been no significant change in perceptions of the U.S. Army culture, and certainly no rise in the perception of increased clan or adhocracy type cultural characteristics. Dr. Pierce states:

> What they really show is that an organization's culture is without a doubt extremely difficult to change especially by superficial changes like *modifying* the OER, changing promotion gates, changing the uniform, changing the length of career courses, sending *some* officers to civilian schools, or creating a mandatory reading list, etc. In other words, our culture is far more persistent than the era of conflict that we find ourselves in.[35]

OFFICER EVALUATION REPORT DATA ANALYSIS

A brief examination of recent Officer Evaluation Report (OER) data supports the findings in the Pierce studies. One of the most powerful influencers of Army culture is the OER. What the cadets listening to Dr. Deresiewicz's speech and preparing to "think for themselves and act on their convictions" in their transition from cadet gray to Army Blue are unaware of and do not fully understand is that the current evaluation and promotion system in the Army drives many leaders to Massengalian tendencies instead of Damonian. Some officers are rewarded on OERs for pleasing higher headquarters, not rocking the boat, and maintaining equilibrium in the system. The OER also sends signals to the rated officer on "what's important in the profession" and what is expected of a leader to achieve professional success.

The current OER plays a highly significant role in how the culture of the officer corps has evolved in the last 25 years. The DA Form 67-9 (Appendix II) that replaced the 67-8 in 1997 was meant to overcome the shortcomings of the old OER, such as overinflation in assessing performance and potential, and to be a developmental tool as well as an evaluation of performance. Unfortunately, the subversive nature of organizational culture helped to undermine the good intentions of the current OER, and it has continued to foster hypercompetitiveness and can inspire a "market culture" over a "clan culture" as highlighted in Dr. Pierce's paper.

The current OER relies on the Rater (supervisor) and Senior Rater (supervisor's supervisor) to assess performance and potential. The front of the OER includes a list of skills, attributes, actions, and "yes/no" criteria of evaluation with an **additional X by areas in each category that highlight the rated officer's strengths.** The yes/no + X method does not adequately communicate to promotion and command-selection boards (or to the officer for developmental purposes) the true strengths (and weaknesses) of the individual officer. More importantly, the front side constitutes a significant signal to officers on "what's important in the profession." The data on the following pages collected from Human Resources Command are a snapshot of recently selected battalion commanders and may provide insight into how the institution rewards skills, attributes, and actions conducive to a hierarchical organization that embraces the old notion of traditional strong-man leadership, control, and stability over an adhocracy culture that rewards adaptive leadership, flexibility, and discretion.

The front side of the OER presents the rater with a list of skills, attributes, and actions, requiring the rater to select ones that "best describe the rated officer."[36] Words matter and have meaning, and the definitions are important in the intellectual discussion of leadership in complex times. It is important to state upfront that all the listed skills, attributes, and actions are salient traits and are necessary characteristics for a well-rounded and balanced leader. For example, there are times when routine problems arise in which the problem definition and solution are clear, and the leader must have the expertise, proficiency, and professional knowledge to provide the group with the right solution from a traditional authority-based leadership style. However, reiterating the themes of the ACC and FM 3.0, the future OE will be one of complexity and ambiguity and, as such, it is harder to develop conceptual skills that the profession should want to nurture and reward throughout the organization to align better with its values and actions.

Figure 2-5 provides the list of Army definitions, which can be found on the front of the OER, to assist in clarifying what the meanings of the words are for raters and rated officers. Highlighted are key terms to explain better the characteristics of the attributes, skills, and actions. The words and phrases that correspond to key concepts and characteristics of traditional authority-based leadership are highlighted in bold. Characteristics associated more prominently with adaptive leadership are italicized:

The following OER data used in the analysis are from a random sample of FY12 combat arms battalion command selectees, who represent the top 15 percent of a given year group cohort and are those most likely to be promoted to colonel and to work on complex strategic issues at the highest levels of the Department of the Army (DA) and the Department of Defense (DoD). They are the group from which the general officers for the next 10-20 years will come. The data may provide some insights into what the profession believes makes a quality leader and potentially identifies the signals it sends and the traits it unconsciously looks for when determining who should be promoted and who should lead the organization at the strategic level into the future.

Using Army definitions, these skills, attributes, and actions can be separated into groups more associated with the traditional notion of leadership ("authority is leadership") or more associated with characteristics found more readily in adaptive leaders. Figure 2-6 also includes the percent of "Xs" marked per grouping on the sample FY12 selected battalion commanders.

```
Attributes:
    MENTAL- possesses desire, will, initiative, and disciple
    PHYSICAL- maintains appropriate level of physical fitness and military bearing
    EMOTIONAL- displays self-control; calm under pressure.
Skills:
    CONCEPTUAL- demonstrates sound judgment, critical/creative thinking, moral
reasoning
    INTERPERSONAL- shows skill with people; coaching, teaching, counseling, motivating,
and empowering
    TECHNICAL- possess the necessary expertise to accomplish all tasks and
functions
    TACTICAL- demonstrates proficiency in required professional knowledge,
judgment, and warfighting
Actions:
    COMMUNICATING - displays good oral, written, and listening skills for individuals
and groups
    DECISIONMAKING - employs sound judgment, logical reasoning, and uses
resources wisely
    MOTIVATING - Inspires, motivates, and guides others toward mission
accomplishment
    PLANNING - develops detailed, executable plans that are feasible, acceptable, and
suitable
    EXECUTING - shows tactical proficiency, meets mission standards, and takes care
of people/resources
ASSESSING - uses after-action [reflection] and evaluation tools to facilitate consistent
improvement
    DEVELOPING - invests adequate time and effort to develop individual
subordinates as leaders
    BUILDING - spends time and resources improving teams, groups, and units; fosters
ethical climate
    LEARNING- seeks self-improvement and organizational growth; envisioning, adapting,
and leading change.
```

Source: DA Form 67-9, Officer Evaluation Report.[37]

Figure 2-5. Army Definitions of Leader Attributes, Skills, and Actions.

Attributes	% X on OER	Skills	% X on OER	Actions	% X on OER
Traditional	15%	Traditional	60%	Traditional	75%
Physical		Technical		Decision-making	
		Tactical		Planning	
				Execution	
				Motivating	
Adaptive	15%	Adaptive	40%	Adaptive	25%
Emotional		Conceptual		Communicating	
		Interpersonal		Assessing	
				Learning	
*Mental, required for Traditional and Adaptive Leadership received				Developing	
70% of the attribute X's				Building	

Source: DA Form 67-9, Attributes, Skills, and Action for FY 12 Command-selectees, sample data from U.S. Army Human Resources Command.[38]

Figure 2-6. OER Attributes, Skills, and Action for FY 12 Command-Selectees.

The first significant divergence in traditional and adaptive leadership characteristics is identified in the "Skills" category. Technical and Tactical skills, which are most closely associated with solving routine (nonadaptive) problems, received 60 percent of the markings, whereas Conceptual (creative thinking) and Interpersonal (coaching, teaching), which are key elements of adaptive leadership, received only 40 percent.

Of most concern were the results of the Actions analysis, which best describes the types of actions leaders take and what is actually seen by subordinates. The actions most closely related to traditional notions of leadership are Planning, Decision-making, Executing, and Motivating; they were identified 75 percent of the time as the actions that most accurately described the leaders selected for command, whereas adaptive actions such as Communicating (listening), Learning (leading change, adapting), and Assessing (diagnosis) received only 25 percent of the Xs. Most revealing was the insignificant percentage of marks for Learning and Assessing—the two most critical actions for adaptive work—their marks making up a combined total of less than 8 percent.

What can this sample tell the Army? We can reasonably interpret this data as saying that the Army selects leaders who value technical and tactical skills over adaptive skills of conceptual and interpersonal abilities. We can infer that the physical component of leadership is considered equally important as emotional intelligence. Finally, being a planner who can lead execution and act decisively for the group is highly favored over a leader who is identified as having the strength to assess (diagnose), communicate (intervene), and learn (lead change).

These percentages suggest that the leaders selected for battalion command are best described by their raters as embodying the traits more closely aligned with the traditional model of leader-follower than with the exemplar of adaptive leadership. As we saw earlier, the former is more likely to struggle in complex environments, while the latter provides the best chance of overcoming adaptive barriers.

Another way to look at the front side of the OER and its use as a tool to communicate to promotion and command boards is to assess it through the lens of cultural gamesmanship. Hypothetically, raters who want to give their star players a "bump" with the senior rater or the board will mark the X that will place the rated officer in the best light, with regard to supporting the profession's basic underlying assumptions. The rater wants to send a signal that this officer embodies the skills, attributes, and actions that we look for in our leaders (not the values we say we want but the values we promote and reward): Mental-Physical, Technical-Tactical, Plan-Decide-Motivate-Execute.

By marking the data on the front side of the OER, as demonstrated in the preceding chart, and then subsequently selecting these officers for promotion/command, the Army's leaders perpetuate the contradiction in the organizational culture. Though most will acknowledge that selection boards place little or no weight on the front side of the OER, it is still a subtle but important data point and a signal, weak as it may be, to the junior officers of what senior leaders think is important and what gets things done.

The backside of the OER is open for subjective review of performance and potential. This includes the "Senior Rater Block Check," applied to field grade officers[39] as a means of identifying the top 49 percent and is a key signal for increased chances of selection for command at the lieutenant colonel level. While a healthy level of internal competition improves overall performance of an organization, the current senior rater blocking system is a primary source of hyper-competitiveness within the officer corps and can feed into a potentially dysfunctional, even toxic, competitive culture within the profession. In the end, a reasonable conclusion can be made that the front side of the current OER reinforces a traditional hierarchical leadership culture, with the back side encouraging a market culture of hard-driving competition, in the face of espoused values calling for adaptive behavior, selfless service, and team play. This simple OER data analysis reinforces the findings of Dr. Pierce, providing yet another data set that calls into question the present alignment of espoused values and basic underlying assumptions in the professional culture of the U.S. Army.

ARMY LEADERSHIP

As the Army works through its adaptive challenge of bringing congruence to its values and practices, it should consider a broader understanding of leadership with regard to leading in the face of complex problems. Developing creative leaders is only one piece of the puzzle. Just because a leader has creative attributes does not mean that he or she understands the tasks of leading change, of mobilizing groups to face reality, and of then guiding them through the problem-solving process. A leader with the ability to be creative can fall victim to the seductive ways of counterfeit leadership. To exercise leadership, it is imperative that the institution teach leaders the skill of leading in the context of adaptive challenges.

Currently, *Field Manual (FM) 6-22, Army Leadership*, defines leadership as "the process of influencing people by providing purpose, direction, and motivation while operating to accomplish the mission and improving the organization."[40] Nowhere does this definition touch on leadership's most challenging function of leading change by mobilizing people to do adaptive work. While providing purpose, direction, and motivation are critical tasks for an organizational authority figure, if a more holistic understanding of leadership is not imbued in the authority figure, these very dimensions can lead to counterfeit leadership. This is because an authority figure may enable an organization to avoid the reality of a changed environment through a false purpose and direction, or misleading motivation.

The types of problematic challenges that leaders at all levels of authority confront in today's OE call for a style of adaptive leadership requiring more than simply purpose, direction, and motivation. Broadening the Leader Core Competencies to include "One who mobilizes—Navigates organizations through adaptive challenges by confronting people with the reality of changed conditions and provides a learning environment for the group to discover and develop adaptive solutions," will help capture the new skills and framework required to be an adaptive leader (see Appendix III).

WHAT'S AT STAKE?

What if the Army chooses not to confront the contradiction between espoused values and its basic underlying assumptions, thus hindering leader development? Or if the organization takes only half steps when full steps are required to fully realize an adaptation? What is at stake is the Army's ability to produce the quantity and quality of adaptive leaders who are creative, imaginative, and innovative and can lead learning organizations on today's competitive battlefield. It is the risk of operating at 80-percent effectiveness as an organization when 90 percent or better can be achieved with cultural realignment. It is the question former Secretary of Defense Gates posed to the Army leadership and was his main worry—"How can the Army break up the institutional concrete, its bureaucratic rigidity in its assignments and promotion processes, in order to retain, challenge, and inspire its best, brightest, and the most-battle tested young officers to lead the service in the future?"[41] What is at stake is an Army that has counterfeit leaders that slip through the cracks and poison junior leaders with toxic leadership. Most important, what is at stake is our ability to win, learn, focus, adapt, and win again.[42]

Retired General Stanley McChrystal, former commander of North Atlantic Treaty Organization (NATO) Forces in Afghanistan, stated during a visit at Harvard University that "the senior guy isn't the expert these days . . . that leadership today must be done by, with, and through others."[43] He gives an example of his grandfather leading in the Army in the mid-20th century. General McChrystal explains how so little had changed between the time his grandfather was a junior leader to when he was a more senior leader. As a result, his grandfather could rely on his past experiences and expertise to lead the group. This was because the problems were not really much different, weapons and tactics were basically unchanged, and he could rely on a repertoire of solutions from his past to solve the problems of the present.

General McChrystal added that the 21st century is much different, that today's senior military leaders must learn from the bottom up and from the group as a whole, as they lead organizations in a world that has changed so drastically from when they were young leaders. He asserts that the expertise no longer lies with seniority or authority, but from within the group, and that the leaders, to be effective, must "have a touch of humility, must stop, listen, and allow others to inform you, and teach you."[44] Leadership in today's rapidly changing environment must reflect the truth that "the wisdom of the whole is greater than the wisdom of one."[45]

Chapter 3 presents an innovative framework for leading groups to overcome adaptive challenges. Its principles, which parallel the themes of General

McChrystal's insights into leadership in the 21st century can, if applied, assist Army leaders in tackling the institutional adaptations it confronts. Adoption of the leadership framework depicted in the next chapter will increase the Army's likelihood of growing adaptive leaders.

ENDNOTES - CHAPTER 2

1. Rollo May, *The Courage to Create*, London, UK: Norton, 1975, p. 27.

2. Robert Gates, Speech to the U.S. Corps of Cadets, United States Military Academy, West Point, NY, March 2011.

3. *Ibid.*

4. *Ibid.*

5. William Deresiewicz, Speech at West Point, Spring 2010.

6. Scott Krawczyk, quoted in *Ibid.*

7. As explained by Deresiewicz.

8. Herbert R. McMaster, posted by Thomas Ricks, "The Best Defense," *Foreign Policy*, April 13, 2010, available from *ricks.foreignpolicy.com/posts/2010/04/13/bg_hr_mcmaster_allow_active_mistakes_but_don_t_tolerate_errors_of_passivity*.

9. As quoted by Daniel P. Bolger, "Zero Defects: Command Climate in First Army, 1944-1945," Fort Leavenworth, KS: L200 Leadership Applied Readings, 2008, p. 51.

10. Robert L. Caslen, Discussion with National Security Fellows, Harvard Kennedy School, March 9, 2011.

11. Bolger, p. 52.

12. Herbert R. McMaster, "Learning from Contemporary Conflicts to Prepare for Future War," *Foreign Policy Research Institute*, October 2008, available from *www.fpri.org/enotes/200810.mcmaster.contemporaryconflictsfuturewar.html*.

13. Raymond Odierno, Discussion with National Security Fellows, Harvard Kennedy School, March 10, 2011.

14. James G. Pierce, *Is the Organizational Culture of the U.S. Army Congruent with the Professional Development of its Senior Level Officer Corps?* Carlisle, PA: Strategic Studies Institute, U.S. Army War College, September 2010, p. iii.

15. Edgar Schein, *The Corporate Culture Survival Guide*, San Francisco, CA: Jossey-Bass Publishers, 1999.

16. Edgar Schein, *Organizational Culture and Leadership: A Dynamic View*, San Francisco, CA: Jossey-Bass Publishers, 1985, p. 18.

17. Edgar Schein, *Organizational Culture and Leadership*, 2d Ed., San Francisco, CA: Jossey-Bass Publishers, 1992.

18. Pierce, p. 12.

19. *Ibid.*, p. xx.

20. *Ibid.*, p. 11.

21. Martin E. Dempsey, "A Campaign of Learning to Achieve Institutional Adaptation," *Army Magazine*, Vol. 60, No. 11, November 2010, p. 34.

22. Martin E. Dempsey, "Leader Development," *Army Magazine*, Vol. 61, No. 2, February 2011, p. 26.

23. The work in the center, the adaptive challenge or the "real" work confronting the group.

24. Pierce, p. xv.

25. *Ibid.*

26. *Ibid.*, p. iv.

27. *Ibid.*, p. 44.

28. *Ibid.*

29. *Ibid.*

30. *Ibid.*, p. xv.

31. *Ibid.*, p. 53.

32. James G. Pierce, email to the author, March 17, 2010.

33. Pierce, p. v.

34. Pierce, email to the author.

35. *Ibid.*

36. Department of the Army Form 67-9, *Officer Evaluation Report*, October 2007.

37. *Ibid.*

38. OER Front-side: Attributes, Skills, and Action for FY 12 Command-selectees, sample data from U.S. Army Human Resources Command.

39. Field Grade officers: commissioned officers in the ranks of major, lieutenant colonel, and colonel.

40. *Field Manual (FM) 6-22, Army Leadership*, Washington, DC: HQ Department of the Army, 2006, pp. 1-2.

41. Gates.

42. Martin E. Dempsey, "Win, Learn, Focus, Adapt, Win Again," *Army Magazine*, Vol. 61, No. 3, March 2011, p. 25.

43. Stanley McChrystal, discussion with National Security Fellows, Center for Public Leadership, Kennedy School of Government, Harvard University, April 25, 2011.

44. *Ibid.*

45. *Ibid.*

CHAPTER 3

REAL LEADERSHIP

Systems are interdependent . . . if you want to keep up with the system, then you have to evolve at least as fast as the system. And if you want to get ahead of the system, you have to evolve at least twice as fast.[1]

General Martin E. Dempsey

Leading a system confronted by a challenge demanding an adaptation requires a leader who can transcend the role of authority. A leader must have the capacity to mobilize people to confront their predicament and solve their problems. The focus of "real leadership," as presented by Harvard Kennedy School professor Dean Williams, is not to get others to follow, but to get people to confront reality and change values, habits, practices, and priorities in order to deal with the *real* threat or opportunity the group faces.

Looking at leadership through a lens provided by Dean Williams reveals a framework for approaching adaptive challenges that the U.S. Army can exploit as it tackles those challenges at the tactical, operational, and strategic levels of war. Williams defines leadership as "a process of mobilizing people to confront and address problematic realities for the purpose of improving the human condition and making progress."[2] Williams posits that notions of leadership based on a model suggesting that "leaders must craft a vision, motivate people through persuasive communication, be an example, and employ a system of punishments and incentives to sustain action, is insufficient for dealing with the complexity of the challenges institutions and communities face in the age of globalization."[3]

He presents an alternative idea of what it means to be a real and responsible leader, one "that does not emphasize the *leader-follower and goal* but the dynamic of *leadership-group and reality*."[4]

Williams defines **an adaptive challenge as a problem that persists even when authority figures and organizations apply best practices and known methods or when standard operating procedures do not solve the problem**. In these cases, the solution to the challenge usually requires a shift in the values and mindsets of the members in the group, making it an adaptive challenge. The "work" or learning people must do to overcome adaptive challenges is simply **adaptive work**, which Williams defines as "the effort that produces the organizational or systemic learning required to tackle tough problems."[5]

COUNTERFEIT LEADERSHIP

Over the last 25 years, the proliferation of leadership and management literature has revolved around the relationship, links, and interdependence of authority, influence, power, and leadership. In addition to providing a framework for the study and application of adaptive leadership, Williams provides insights into false or maladaptive leadership, which he describes as "counterfeit leadership."[6] Distinguishing between "real" leadership and "counterfeit" leadership is essential to understanding and applying leadership in the face of adaptive challenges. Acknowledging the dichotomy of styles is fundamental to the study and understanding of adaptive leadership.

Williams, in his book *Real Leadership*, and, even more so, Ron Heifetz in *Leadership Without Easy Answers*, go to great lengths to uncover and explain the relationship of authority and dominance in social groups and how it affects the exercise of leadership. Both authors link human nature to our evolutionary instincts to seek protection and equilibrium within our social systems by looking to authority for answers to problems confronting the group. Linked to this natural desire to look to authority for solutions is our primal instinct that when in positions of authority, we tend to *assume the role* as the Alpha in the group (even if it is not our nature to do so). Basic primate groups look to the Alpha for this protection. The Alpha uses dominance in his role as the designated authority and leads the group through this means. Reflecting on former groups and organizations, most of us can think of times when dominant personalities have used their formal or informal authority to run the organization through dominance, fear, coercion, and ridicule.

The U.S. Army espouses the need for leaders who can lead change, work collaboratively, and run organizations that embody the ideals of operational adaptability. Leadership through dominance will not produce the climate in an organization conducive to fostering these essential skills and attributes. As such, we would like to believe that the Army has a culture that winnows out smothering leaders through a command-selection process that withholds rewarding the behavior of those who use dominance to lead—in other words, a system that avoids putting counterfeit leaders in command.

Unfortunately, as late as January 2011, the Army had to relieve a brigade commander "due to a loss of confidence in his ability to command."[7] The investigation described a commander who demonstrated counterfeit leadership, what the journalist Jeff Gould described as "toxic leadership." The Army's inves-

85

tigating officer stated that the commander who was relieved "demonstrates arrogance, deception, and threatening behavior," producing a "command climate that was 'at best ineffective, and at worst toxic'."[8] How did a leader with these traits rise to one of the highest tactical commands in the U.S. Army? What behavior was overlooked or even rewarded over his 22 years of service before he was handpicked by one of the most competitive selection boards in the Army to command the organization's centerpiece maneuver formation—the Brigade Combat Team manned by 4,000 U.S. Soldiers? These are important questions for leaders managing the Army's culture.

Persons in positions of authority, which in the U.S. Army are "the commanders," who lack the ability to reflect, to stand back from the fray to take a large view of the system, and who cannot **learn**, are prone to providing counterfeit leadership. The avoidance of counterfeit leadership can thus be a significant waypoint in an authority figure's approach to moving an organization forward. As members of social systems, humans depend on authority to provide order and structure. This dependency can cause a group to have false expectations of the authority figure, i.e., expecting the commander to provide all the answers. This is because, as Riley Sinder and Ron Heifetz of Harvard's Kennedy School have stated: "[I]n our everyday language, we equate leadership with authority."[9] Because we so commonly equate leadership with authority, we fail to see the obstacles to leadership that come with authority itself. "Having authority brings not only resources to bear but also serious constraints on the exercise of leadership."[10] Unlike leadership, which is a choice, authority is a service. Authority provides order, protection, direction, and accountability. In times of high stress, an overreliance on authority to bring equilibrium to the system or group can bring a false sense of security. Authority is frequently embraced by the group, and in return it feeds the group's hunger for equilibrium. It does this by providing simple technical solutions to complex problems, which allow the group to avoid the work required to confront the need to adapt.[11] This dynamic results in counterfeit leadership.

REALITY AND LEADERSHIP

The primary duty of a leader is to **help the group face the reality** of their problematic challenges. The leader must take on this responsibility because, if not, the group will engage in work avoidance—avoiding the reality of the changed conditions—because reality in adaptive scenarios will force group members to confront their countervailing values and deeply rooted beliefs, thereby causing disequilibrium in the system. Avoiding reality by attacking symptoms and making excuses allows people to "distance themselves from the responsibility of the real issue," Williams says.[12] It is human nature for people to avoid acknowledging a changed environment in an effort to maintain equilibrium within their systems. Facing the reality of a changed world is frightening, but when confronted by an adaptive challenge, the leader must be willing to take the responsibility to impose disequilibrium on the system, no matter how painful or disturbing it might be to the members of the organization. This must be done before the group can begin adaptive work. It is in the state of disequilibrium where real learning takes place.[13]

A NEW FRAMEWORK FOR ADAPTIVE LEADERSHIP

The exercise of leadership is discretionary, involving a choice as to whether to take responsibility and "respond to problems, dysfunctions, and tremendous opportunities that emerge in our organizations."[14] Williams presents the following framework for exercising leadership. To get people to face their adaptive challenges, a leader must be responsible for the following:

1. The diagnostic process, determining the precise nature of the problem and assessing the group's readiness to confront the problem.

2. Manage the problem-solving process as *an inclusive process by the group* as it works through conflicts in values and priorities and embraces new practices that bring resolution to problematic situations and open up pathways for genuine progress.

3. Conceiving of oneself as an instrument of power, that is, achieving awareness of how one's power — authority, presence, and interventions — affect the thinking and actions of others as they tackle their challenges is essential to success. "Ultimately, one's power and authority must not be used to get others to follow, but to get the group to confront reality, so necessary for adaptive work."[15]

As depicted in Figure 3-1, this framework requires the leader, who is an integral part of the system, to pull himself/herself out of the system, to sense the environment by listening to all factions, while diagnosing the true nature of the challenge. The leader then develops an intervention strategy and reenters the system to confront the group with the reality of the changed environment. Then, working through the system, the leader helps the group do adaptive work. The leader manages the learning process by preventing the group from engaging in work avoidance and keeps them focused on the work at the center. The leader ensures that the right level of disequilibrium is present for real learning to take place and helps the system achieve the aspiration to solve its adaptive challenge.

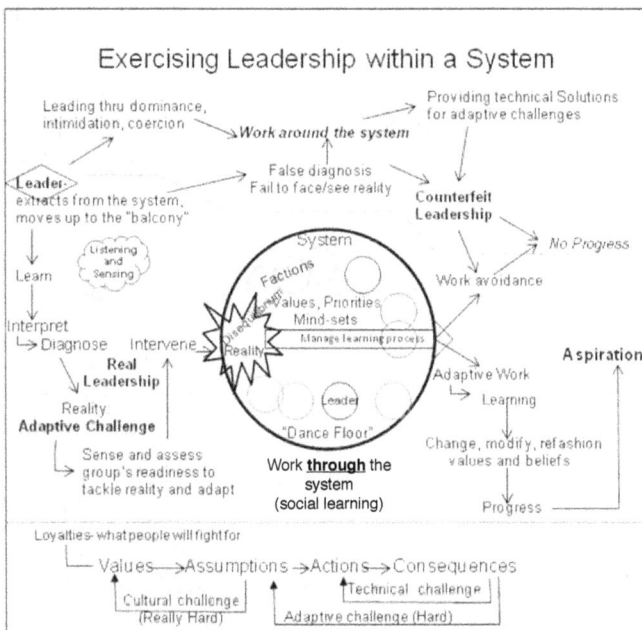

Source: Using concepts from Dean Williams, Ronald Heifetz, and Marty Linsky, the author developed the visual depiction of the theory of exercising leadership in a system as presented by Dean Williams in MLD-201: Exercising Leadership: The Politics of Change, Harvard Kennedy School, January-April 2011.[16]

Figure 3-1. Author's Conceptual Depiction of Dean Williams's Exercising Leadership within a System.

Throughout the process, the leader must be aware of him- or herself, assuring that he/she does not allow the group to seduce him/her to become the technical solution provider. The leader's self-awareness is essential to ensuring that he/she does not divert the group through counterfeit leadership by using power to get others to follow, instead of using the authority and power to get the group to face reality, prevent work avoidance, and adapt to solve the problem. The leader must also ensure that he/she is not "killed off" (marginalized, undermined, ignored, or relieved/fired) by the group or by those above, either as a result of creating too much disequilibrium or by not meeting expectations of the group to bring equilibrium back to the system.

DIAGNOSING THE PROBLEM

The first and perhaps the most critical step in the framework is properly diagnosing the problem confronting the group. Dean Williams's mentor Ron Heifetz posits a very similar approach to leadership. Heifetz's model for determining the nature of the challenge helps clarify the nature of the challenge. He distinguishes between three types of problems that face organization (see Figure 3-2). In each one, depending on the degree of complexity, different demands are made on the authority figure and the group; as the complexity increases, the requirement to exercise adaptive leadership increases.

Situation	Problem Definition	Solution and implementation	Primary locus of responsibility for the work	Kind of work
Type I	Clear	Clear	Leader	Technical
Type II	Clear	Requires Learning	Leader and Group	Technical and Adaptive
Type III	Requires Learning	Requires Learning	Group>Leader	Adaptive

Source: Heifetz, *Leadership Without Easy Answers*, Cambridge, MA: Harvard University, 1994, p. 76.[17]

Figure 3-2. Heifetz Table of Situation Types.

The leadership required to handle Type I challenges requires only technical solutions. These problems are routine and the solutions are readily available. Authority figures rely on their education, training, and experience to solve the group's challenges. When confronted by Type II and III challenges, authority figures frequently find themselves at a fork in the road, where one path leads to real leadership and the other to counterfeit leadership. This is because authority figures often insist on being the problem solver, providing a technical solution when in actuality, adaptive solutions require the group to find the real solution.

For the U.S. military at the strategic level, the complexities of the problem set will always present Army leaders with Type III challenges, requiring a more inclusive notion of leadership, as described by Williams and Heifetz. Likewise, in the current operational environment (OE), our tactical-level leaders who used to live in a Type I world of linear tactics, routine garrison management, and personnel actions, now face the complexity and ambiguity found in Type III problem sets, which require different skill sets. Leaders without the skills to adapt, when presented with a Type II or III problem, may default to counterfeit leadership. They will choose the path of work avoidance mechanisms such as "holding onto past assumptions, blaming authority, scapegoating, externalizing the enemy, denying the problems, jumping to conclusions, or finding a distracting issue [that] may restore stability and feel less stressful than facing and taking responsibility for a complex challenge."[18]

Perhaps most dangerous is the charismatic figure who employs work-avoidance mechanisms while re-establishing equilibrium in the system through persuasive and motivating rhetoric to help the organization avoid the reality of a changing environment.

> Instead of generating creativity and responsibility, charismatic authority can generate a mindless following or devolve into bureaucratic institutions that rely on central planning and control. Creativity is stimulated by engaging with one's environment, but the skills of sensing local environments become dulled as people fasten their gaze on the charismatic figure or the chain of command for direction. Focused upward, people lose touch with their communities, markets, and personal resources.[19]

During the diagnostic process, the leader must leave the "dance floor" (See Figure 3-1) and go up to "the balcony" for a clearer perspective, through which he can make an accurate and honest assessment of the challenge confronting the organization.[20] Once the leader diagnoses the challenge as either a Type II or Type III, the leader must assess the group's preparedness, its ability to confront reality, and its willingness to question assumptions and stimulate growth. After this initial analysis, the leader designs an intervention to move the group toward adaptive work (learning). To do this effectively, the leader must accurately diagnose the domain, or nature, of the adaptive challenge. There are six domains suggested by Williams in which most Type III (adaptive) challenges can be categorized. Clearly diagnosing the nature of the adaptive challenge will assist the leader in developing a strategy to intervene and then manage the problem-solving process of the group. Williams's six domains of adaptive challenges include:

1. **The activist challenge**: Calling attention to a contradiction in values;

2. **The development challenge**: Cultivating the latent capabilities needed to progress;

3. **The transition challenge**: Moving from one system of values to another;

4. **The maintenance challenge**: Protecting/sustaining what is essential during hard times;

5. **The creative challenge**: Doing what has never been done before; and,

6. **The crisis challenge**: Leading in a period of extreme danger.[21]

When confronted by an **activist challenge**, the leadership work is to provoke and evoke, to persuade the group to face certain realities it would prefer to avoid. The process includes both inspiring people with a unifying purpose and calling attention to the contradiction between what the group espouses and what it actually does.[22]

When faced with a **development challenge**, the leadership task should orchestrate a learning process through designed experimentation that cultivates the group's latent capabilities.[23] To ensure the growth or even survival of the organization, it must build new capabilities, competencies, practices, and processes.[24]

Transition challenges emerge when some values and mindsets are no longer useful in addressing challenges of the organization due to a shift in the dynamics of the environment or the emergence of a new threat or opportunity. In this case, the leadership task is to help the group refashion values, loyalties, and mindsets.[25]

In a **maintenance challenge**, leading change is not the focus; rather, the ability to protect essential resources, maintain core values, and keep the enterprise from falling apart becomes the new preoccupation. Ignoring a threat to the group will not resolve the issue; leaders must mobilize the group to overcome dysfunctional practices and return to core values.[26]

When a group is confronted by what appears to be an insurmountable barrier that best practices cannot overcome, then it faces a **creative challenge**. Breaking the current paradigm and advancing to a new one requires people to create.[27] Unlike a development challenge, a creative challenge requires a significant break with the past and an unconstrained leap into the future.[28]

The **crisis challenge** is an explosive situation in which survival is at stake and urgent action is required. The group is under attack and its accrued value is at risk.[29] (A deeper and more thorough explanation of the principles in Williams's new framework for exercising leadership can be found in outline form in Appendix I.)

The essence of leading in these complex domains is the ability to remain flexible. Often adaptive challenges facing organizations will have elements of multiple domains and require agility of mind and effort by the leader to mobilize the group to address the reality of the day. An organization can face multiple adaptive challenges simultaneously, which further complicates the already complex nature of the challenges faced by the group.

General Odierno and the next generation of Army leaders will continue to be faced with multiple adaptive challenges. General Dempsey's intervention in the form of the "Campaign of Learning" was, in and of itself, a means to tackle the wide range of adaptive challenges confronting the Army after a decade of conflict and high operational tempo. Within this campaign, General Dempsey's effort to reexamine what it means to be a profession was and remains an essential maintenance challenge as the Army struggles to balance and maintain core values. The doctrinal shift from command and control to mission command is an example of a transition challenge, and instituting a new conceptual foundation is a development challenge. All the while, the potential of a crisis challenge looms in this complex and dangerous world. However, it is leader development, which General Dempsey identified as "job number one," that continues to present the Army with its most important and difficult adaptive challenge.[30] Actually, however, the nature of the challenge, because it will meet resistance as a result of strongly held practices and assumptions, makes it an *activist challenge.*

ENDNOTES - CHAPTER 3

1. Martin E. Dempsey, Leader and Concept Development Speech, Association of the U.S. Army Winter Symposium and Exposition, Fort Lauderdale, FL, 2010.

2. Dean Williams, HKS Course MLD-201, Exercising Leadership: The Politics of Change, Harvard Kennedy School MLD-201, January 24, 2011.

3. Dean Williams, *Real Leadership: Helping People and Organizations Face Their Toughest Challenges*, San Francisco, CA: Berrett-Koehler Publishers, 2005, p. 5.

4. *Ibid.*

5. *Ibid.*, p. 7.

6. *Ibid.*, p. 14.

7. Jeff Gould, "Germany-based colonel relieved of duty," *Army Times*, March 6, 2011, p. 32.

8. *Ibid.*

9. Riley Sinder and Dean Williams, "Leadership Styles," *Clinical Laboratory Management*, Washington, DC: ASM Press, 2004, p. 209.

10. Ronald A. Heifetz, *Leadership Without Easy Answers*, Cambridge, MA: Harvard University Press, 1994, p. 50.

11. Sinder and Williams, p. 209.

12. *Ibid.*, p. 11.

13. Williams, "Exercising Leadership."

14. Williams, *Real Leadership*, p. 12.

15. *Ibid.*, p. 13.

16. Using concepts from Dean Williams, Ronald Heifetz, and Marty Linsky, the author developed the visual depiction of the theory of exercising leadership in a system as presented by Williams, "Exercising Leadership."

17. Heifetz, p. 76.

18. *Ibid.*, p. 77.

19. *Ibid.*, p. 66.

20. Ronald A. Heifetz and Marty Linsky, *Leadership on the Line: Staying Alive through the Dangers of Leading*, Boston, MA: Harvard Business School Press, 2002, p. 53.

21. Williams, *Real Leadership*, p. 57.

22. *Ibid.*, p. 231.

23. *Ibid.*, p. 89.

24. *Ibid.*, p. 90.

25. *Ibid.*, p. 115.

26. *Ibid.*, p. 142.

27. *Ibid.*, p. 163.

28. *Ibid.*, p. 165.

29. *Ibid.*, p. 189.

30. Martin E. Dempsey, "Leader Development," *Army Magazine*, Vol. 61, No. 2, February 2011, p. 26.

CHAPTER 4

THE REAL WORK AHEAD

The tendency of any big bureaucracy is to revert to business as usual at the first opportunity — for the military, that opportunity is, if not peacetime, the unwinding of sustained combat.[1]

— Robert Gates, Secretary of Defense
Speech at West Point, February 2011

The U.S. Army does many things right — its leaders and Soldiers have fought and sacrificed valiantly in service to the Nation; it cares for its wounded, honors its dead, and supports its families. However, despite its many successes, the U.S. Army can never be satisfied with the status quo. As a learning organization, the Army must view self-improvement as a sacred professional obligation. The profession of arms enjoys the trust and confidence of the American people because it self-polices. As General Martin Dempsey stated, "During the last 9 years of conflict, our Army has shown itself to be both introspective about its performance and adaptive to the lessons it has learned. Nevertheless, we want to formalize the effort to learn."[2] Secretary Gates highlighted this impulse with the observation that "the Army's ability to adapt allowed it to pull Iraq from the brink of chaos in 2007, and in 2010 roll back the Taliban in Afghanistan." He then quoted the words of General Peter Chiarelli, Vice Chief of Staff of the Army: "It is important that the hard-fought lessons of Iraq and Afghanistan are not merely 'observed' but truly 'learned' — incorporated into the service's DNA."[3]

For the U.S. Army, the most pressing adaptive challenge is overcoming the contradiction between espoused values and actual behavior with regard to Army leader development. We must revisit the work in the center of the contradiction, "aligning and connecting our leader development programs and policies (basic underlying assumptions) with our conceptual foundation and doctrinal changes such as mission command (espoused values)."[4] This effort is the most critical adaptation we can make within the Army's campaign of learning. This alignment is critical, because it is necessary in order to develop adaptive leaders, which are required to lead learning organizations. At first glance, this proposed adaptation appears to be in the transition domain. However, the fact is that this complex challenge is rooted in an almost intractable organizational culture whose nature demands a radical professional airing. Hence, we can accurately say that the needed adaptation occupies the activist domain as well as the transitional: "The first challenge of leadership [in an activist challenge] is to get people to wake up to the fact there is a problem—that the group is avoiding some aspect of reality, ignoring a threat, or missing an opportunity."[5]

There are thus times when leaders must take an activist role in leading change—to get people to wake up and face the problem or, in this case, to seize an opportunity. As Williams states, "Often the problem is embedded in the people's values and behavior. People might espouse one view but act in ways that are not consistent with that view. The leadership task in an activist challenge is to call attention to the contradiction in values and intervene to disrupt the thinking and patterns of behavior that allow the people to persist in avoiding the reality of their condition."[6]

Symptoms of an activist challenge for any group include:

- "Some enduring behaviors, values, and/or *practices have become corrosive and dysfunctional* and serve to undermine the long-term integrity and survival of the group.
- An opportunity presents itself that can lead to great benefit and progress for the group, *but no one is seriously considering it.*
- Danger is looming due to an internal or external threat, *and the group is not doing anything about it*"[7] (Italics added by the author).

In diagnosing an activist challenge, the leader must examine the condition of the people facing the challenge, the barrier that impedes progress, and the promise or aspiration on the other side of the barrier. Generally speaking, the *people* are unwilling to change their values or thinking to accommodate some aspect of reality.[8] The people are in denial, resistant to change, and comfortable where they are. The *barrier* in an activist challenge is the people's resistance. Individual thinking is trapped by the prevailing group paradigm.[9] The *promise* is that if people can face the problem and seriously consider the data that they have neglected or denied, then a new opportunity for progress can open. In these cases, fundamentally, the leader wants people to learn, to learn about the problem, to learn *how they contribute to the problem,* and what can be done to solve the problem.[10] This learning may mean swimming upstream against the current of organizational culture. Or, to switch metaphors, the leader must "burrow down into people's underlying assumptions and deeply held beliefs to ascertain why the problem persists and what can be done about it."[11]

As the Army reflects on the state of the profession today, it is clear that certain behaviors and practices inside the organization have become corrosive and dysfunctional and could potentially undermine the long-term integrity of the profession. General Dempsey identified reality, the dangers confronting the profession internally and externally in the operational environment (OE). His intervention was manifested in the Army's "Campaign of Learning," which challenged its members to use their imaginations to improve the profession. During the campaign of learning, there is opportunity for institutional growth and for real change to move the organization forward. Is the *group* ready for the adaptive work? Is the profession ready to face the reality that it must abandon some of its subconscious insistence on control and stability, that it must push boundaries in order to implement changes that will cause disequilibrium and rock the boat? Is it prepared for the discomfort of having its basic underlying assumptions questioned and challenged in an effort to achieve cultural adaptation? As Dr. James Pierce stated, "A real culture change will require a substantial reengineering of the way we do business . . . for example, making our vocation a true life-long profession, and, of course, a complete overhaul of the evaluation and promotion system."[12] As we recall from an earlier chapter, Dr. Pierce's study indicates that the promotion system may itself be the source of a toxic competitive culture.

POTENTIAL BARRIERS

Ironically, the greatest barrier to the Army's adaptive work is the leaders in the organization itself, specifically, the field grade officers who have the greatest impact on the day-to-day operations of organizations and the greatest influence on the mentorship and molding of the junior leaders in the organization. It is the power of organizational culture and its resistance to change that will challenge these leaders to fully embrace the adaptations required to alter how the Army fundamentally does business. Field grade officers, primarily majors and lieutenant colonels with 10-20 years of service, are both fully immersed in and committed to the Army culture and remain professionally competitive as they strive to embody the professional ideals and compete for senior command selection. This combination of cultural influence and professional competitiveness makes them vulnerable to risk aversion and work avoidance with regard to the effort associated with adaptation.

What Williams describes as **work avoidance**—a response to disequilibrium in which individuals or groups eschew work that challenges their values or beliefs[13]—will manifest itself through these leaders. The campaign of learning threatens some deeply held beliefs, conscious and subconscious, causing pushback against the learning process. Leaders will find "more pressing" issues to occupy their time rather than deal with the real work at hand. The leadership challenge is to get people to confront the gap between their aspiration (growing adaptive leaders) and what they are willing to contribute or sacrifice in order to fulfill that aspiration (willingness to take prudent risk, to make oneself vulnerable in decentralized operations, to allow junior leaders to develop by underwriting mistakes made by subordinates in the name of experimentation and learning).[14]

In a "get the job done" culture, leaders unwilling to risk failure will employ work-avoidance mechanisms that will subvert the institutional adaptations sought in the campaign of learning. Indicators that leaders are avoiding adaptive work include: leaders who doggedly hold onto outdated assumptions to remain in their comfort zone, leaders who lead through dominance and control, leaders who repeatedly assign provably bankrupt tasks to the group and get them to follow, and leaders who look up the chain of command for solutions.[15]

The greatest pitfall in the get-it-done culture is for the members of the organization to look to its authority figures to provide the solutions. Unfortunately, many will look to General Odierno and other senior leaders for *the answers* to the challenges, to sate their hunger, and to cure their ailments. As the heat goes up with the introduction of disequilibrium, those less adaptive will look to senior leaders (authority) to provide simple answers that will bring equilibrium back to the system—but will also snuff out learning, change, and required adaptation. They will look to the senior leaders' wisdom to provide solutions, when, as Williams states, real leadership wisdom "requires pursuing truth with fervor and passion, being sensitive to the context in which the problem resides, and holding the question in each context, 'What will make our work worthwhile—to our lives and the lives of others?'"[16] Moreover, the real leadership wisdom that Williams advocates is not that of an all-knowing divine; it is the wisdom to "discern which values to promote and protect, and which values should be challenged or changed."[17] The adaptive work, however, is the responsibility of the entire group; it is for all the members of the profession to engage the uncertainty, to be creative and come up with innovative solutions to align values with practice.

CONCLUSION

Despite the present superior comparative strength in U.S. military capabilities, a failure of imagination and inability to achieve operational adaptability could become the U.S. Army's Achilles heel. The Army's true strength must become its intellectual approach to future conflict, where the Army not only has the ability to out-fight the enemy, but to out-think him, to adapt faster and maintain the initiative. During an address at the Joint Warfare Staff College in February 2010, Admiral Eric Olson, Commander, Special Operations Command, stated it most succinctly, "We are not going to fight our way out of these conflicts; we are going to have to think our way out."[18]

Authority, like command, is bestowed upon an individual to maintain order and accountability. Leadership in any given instance offers a choice. It invites potential leaders to assume responsibility. As members of a profession, officers collectively must ensure that the entire organization assumes its responsibility and makes the choice to create learning organizations. These organizations, led by leaders who exercise real adaptive leadership, will ensure that the Army achieves operational adaptability to promote peace or, when called on, to win the Nation's wars.

The Army's "Campaign of Learning" sets conditions for a continuum of learning across the Army that will result in a paradigm shift in the approach to institutional adaptation.[19] Using a more holistic approach to adaptive leadership through frameworks like the one presented by Dean Williams, and confronting the contradictions in the professional culture, are essential to the Army's ability to adapt, which is "an institutional imperative."[20] General Paul Gorman, Commander in Chief, U.S. Southern Command from 1983 to 1985, once said that doctrine and training will prepare us for what lies ahead only if "forceful effective ideas on how to fight *pervade the force*."[21] The Army is fighting effectively today in Iraq and Afghanistan — decentralized, allowing commanders to use discretion and take initiative. However, as General Dempsey has said, "These principles have not yet been made institutional in our doctrine and in our training." As such, they do not pervade the force, nor our culture. "Until they do — until they drive our leader development . . . we [the Army] cannot consider ourselves ready, and we should not consider ourselves sufficiently adaptable."[22]

ENDNOTES - CHAPTER 4

1. Martin E. Dempsey, "Leader Development," *Army Magazine*, Vol. 61, No. 2, February 2011, p. 26.

2. Martin E. Dempsey, "A Campaign of Learning to Achieve Institutional Adaptation," *Army Magazine*, Vol. 60, No. 11, November 2010, p. 34.

3. Robert Gates, Speech to the U.S. Corps of Cadets, United States Military Academy, West Point, NY, March 2011.

4. Martin E. Dempsey, "Leader Development," *Army Magazine*, Vol. 61, No. 2, February 2011, p. 26.

5. Dean Williams, *Real Leadership: Helping People and Organizations Face Their Toughest Challenges*, San Francisco, CA: Berrett-Koehler Publishers, 2005, p. 59.

6. *Ibid.*

7. *Ibid.*, p. 62.

8. *Ibid.*

9. *Ibid.*

10. *Ibid.*

11. *Ibid.*

12. Dr. James G. Pierce, email to the author, March 17, 2010.

13. Dean Williams, HKS Course MLD-201, Exercising Leadership: The Politics of Change, Harvard Kennedy School MLD-201, class January 24, 2011.

14. Williams, *Real Leadership*, p. 48.

15. *Ibid.*, p. 29.

16. *Ibid.*, p. 8.

17. *Ibid.*

18. Eric T. Olson, Lecture at the Joint Forces Staff College, February 2010.

19. Martin E. Dempsey, "Driving Change Through a Campaign of Learning," *Army Magazine*, Vol. 60, No. 10, October 2010, p. 66.

20. Martin E. Dempsey, "Concepts Matter," *Army Magazine*, Vol. 60, No. 12, December 2010, p. 40.

21. Martin E. Dempsey, "Mission Command," *Army Magazine*, Vol. 61, No. 11, January 2011, p. 44.

22. *Ibid.*

APPENDIX I

A NEW FRAMEWORK FOR LEADERSHIP

Dean Williams
Center for Public Leadership
Kennedy School of Government
Harvard University

What Does it Mean to be a Leader?

1. Leadership is the *activity* of *mobilizing* people to *confront and address problematic realities*, engage in learning, and create what is needed to *improve the human condition* or make things better.

2. For every group, there are practices or problems (internal or external) that are impediments to progress. These impediments are adaptive challenges that must be productively resolved for the group to advance. To successfully address an impediment, the exercise of leadership is needed to design and orchestrate a series of experiments to discover what works or what is missing and what must be modified in the group's values, practices and priorities.

3. Our notion of leadership is significantly different from traditional notions of leadership, which over-emphasize the practices of "giving people answers," "gaining followers," "showing the way forward," and "getting people to do what you want them to do." Of course, there are times when giving answers, showing the way forward, and motivating people are important, but the essence of adaptive leadership is in giving the problem-solving work back to the people by getting them to face reality, learn, discover, solve problems, and take responsibility for the work they must do to generate real progress.

4. We distinguish leadership from authority. Most people collapse the two, believing they are the same. We say that authority is a position or function in a group, while leadership is an activity that can be done with or without authority.

5. Authority generally supports the implementation of *technical work* while leadership orchestrates the process of *adaptive work*—a process in which groups modify their thinking and values and make significant adjustments in their behavior to accommodate the reality of a changed context produced by threats and dangers or the emergence of new opportunities and possibilities. Technical problem-solving generally does not require a shift in how people think and behave, while adaptive problem solving does.

6. All systems (groups and institutions) must strengthen their adaptive capacity in order to survive. Evolution is the natural process for unfolding the adaptive capacity of an entity or a system. But the evolutionary process takes a long time and is also inefficient and deficient, because it generates too many losers, produces too much waste, and promotes survival of the fittest. Since the evolutionary process takes too long, systems often die off or collapse because they cannot generate a successful adaptive response to cope with changed conditions fast enough. Leadership is the process of intervening in the system to "punctuate the equilibrium" or "disturb the drift" of the group in order to stimulate problem-solving and change at a faster rate than evolution provides.

7. The intervention process of leadership gets people to attend to what is flawed, broken, or deficient in their thinking, values, norms, or shared patterns of operation. It seeks to put enough reality in the lap of the group so that threats can be acknowledged, flaws fixed, change pursued, and progress generated.

8. Progress, according to the dictionary, means a move toward a "higher or better state."

9. Within any human, organizational, social, political, or economic system, there may be differences of opinion about what is meant by "higher" or "better." Therefore, the first adaptive challenge of leadership will be to work with the group to generate a shared perspective on higher and better that takes into consideration the reality of the group's context, the people's aspirations, and the threats and opportunities before the group.

10. The notion of progress — or what it means to move toward higher and better — should be subject to testing by many factions so that it does not become a narrow agenda of a single faction, and thereby lead to the alienation, attack, or harm of others. Testing what progress means to a group on an ongoing basis should include demanding ethical and moral standards and considerations to ensure that short-term, parochial self-interest does not override the sustainable well-being of the whole, as happened with the U.S. banking system and contributed to the recent financial crisis.

11. The key driver of progress lies in increasing the system's capacity to solve complex problems faster so that a successful adaptation in the system will result — thus minimizing the danger and threat to the system and ensuring it has the capacity to deliver on its promise.

12. Movement toward a higher or better stage by any group requires added investment in time, resources, and energy, and an increase in coordinated effort to generate values, practices, systems, and structures that can produce enough adaptive work so progress unfolds with minimal waste and casualties.

13. This process of producing progress through adaptive work requires examining the prevailing assumptions and deep beliefs people hold concerning key human pursuits. For example, the prevailing belief among many groups is that democracy and the free market are the vehicles to get to higher and better. Another belief is that military interventions on behalf of national interests or the replacement of one regime with another can lead to higher and better. But, as we have discovered through the financial meltdown in the United States and elsewhere, democracy and a free market without enlightened rules, or indiscriminate strategic military interventions, cannot produce spontaneously or by design enough of the progress that the 21st century will require for survival of the valuable parts of civilization.

14. Therefore, it is imperative that leadership be exercised to orchestrate adaptations that produce enough learning, problem-solving, and discoveries in groups to fix deficiencies and faults in human theories, plans of actions, implementation strategies, and coordination mechanisms between factions so that genuine progress eventuates.

Diagnostic Work, Social Learning, and Adaptive Work.

1. Given the threats and dangers in an ever-changing world, complex and challenging problems exist for all institutions, groups, organizations, and societies that expose weaknesses and deficiencies in any human system. We call these problems adaptive challenges.

2. There are different kinds of challenges for every group. The kind of challenge the people face should

be determined by what's at stake if the people persist in their current course. In other words, what are the threats to the system that necessitate people to modify some aspect of their habits, thinking, or priorities?

3. There is an activist challenge — getting people to face and consider something they have been refusing to face or consider; a development challenge — bringing forth latent capabilities in order to respond to the threats and take advantage of opportunities; a transition challenge — shifting the values of the group from one set to another; a maintenance challenge — preserving essential parts of group values in the face of peril or threat; a creative challenge — doing something that has never been done before; and a crisis challenge — attending to ticking time-bombs that threaten to destroy much of the value that has been amassed.

4. Leadership begins with thorough diagnostic work to discover the essence of the adaptive challenge facing the group. However, with tough problematic realities and demanding adaptive challenges, it is not always easy to frame them or put labels on them in a way that is accurate or useful. Often what we describe as the problem is nothing more than a symptom. Adaptive problems are usually messes that are extremely complex and systemic in nature and require a significant degree of diagnostic work to figure out the real issues and the values that perpetuate them.

5. The diagnostic of leadership seeks to help people, according to Plato, distinguish essence from appearance and shadows from reality. In other words, leadership seeks to take a group beyond surface phenomena, myth, and superstition (anything believed to be true that is not) and facilitate a deeper learning pertaining to the assumptions, beliefs, and values people hold with regard to their condition and interpretation of reality.

6. In doing good diagnosis, a leader might ask: 1) What are the internal and external dangers and threats to the system? 2) Do people see and agree on these dangers and threats? 3) How do people interpret the reasons for these dangers and threats? and 4) What values, mindsets, and priorities need to be promoted, changed, discarded or modified in order to address the challenge?

7. There are different kinds of problems facing any group. There are adaptive problems that the people have failed to anticipate because they have no prior experience and therefore do not recognize the phenomenon that indicates danger is in their midst or on the horizon.

8. There are adaptive problems that people recognize but are complacent in addressing because they feel that to take steps to address the problem would require too much of a sacrifice in the present.

9. There are also adaptive problems that people recognize, but their interpretations are incomplete or biased, and they impute blame to others and fail to acknowledge that their own values and priorities are also contributing to the problem.

10. Diagnostic work must consider the systemic dynamics that surround and perpetuate an adaptive problem, because often a problematic concern resides in the *relationship* of the various factions. In other words, there is something deficient or maladaptive in the relationship between the factions that serves to threaten the whole. The framing of the adaptive challenge must therefore include the relationship of the factions.

11. Therefore, a primary function of leadership, in the context of an adaptive challenge, is to diagnose: 1) what the factions are; 2) how the different factions in-

terpret the challenge; 3) how each faction contributes to the problem; and, 4) what values and practices in the relationship between the factions perpetuate the problem.

12. In doing diagnostic work, the leader might ask, "What is unresolved, that if it was resolved, would make a difference for the system? Another important diagnostic question is: "What values do the people hold that they consider more important than facing the problem or making genuine progress?" These questions point at the work-avoidance dynamics in the system and help explain the reasons for the work avoidance.

13. An important facet of diagnostic work is to generate learning and discoveries about the system and the adaptive challenge through experimentation. Discoveries about the reality of the challenge or the group predicament can only unfold and be sustained when deep social learning takes place. Social learning and institutional learning require multiple groups and factions to engage the competing perspectives of the other to enrich their understanding of their condition.

14. To orchestrate a deep social learning, there generally must be some perturbance to the system to generate enough *disequilibrium* that allows for high levels of engagement of competing perspectives. The disequilibrium produces tension as people wrestle with what is essential and what is expendable. It is a conflictual process associated with the discord and the pain associated with a group seeking to make adjustments in their habits and practices and deal with the losses that must be sustained if they are to move forward.

15. Social learning (change, transition, and development) must be a carefully paced process. As Plato's

analogy of the Cave illustrates, you cannot overwhelm the group, or it will turn against you. Helping people sustain losses or let go of maladaptive practices, values, and perspectives is an essential part of the learning process and requires sensitivity, pacing, and a large dose of compassion.

Group Dynamics and Social Systems.

1. Leadership is more than getting a single group or team performing better. It generally requires intervening in a complex system of varying factions and interests. To exercise leadership, one must have an appreciation for the systemic dynamics, which include the unconscious forces that shape and influence the behavior of the individual parts and the collective whole. Groups, organizations, communities, and nations are more than the sum of the individual parts and have a life of their own. They are social systems with multiple interacting components that affect each and every part.

2. As it pertains to a tough leadership challenge, the leadership task is to figure out how individuals act out factional values, loyalties, unresolved concerns, and fantasies, and perpetuate the situation of irresolution. Rather than conduct a psychological analysis of the individuals to figure out why something is broken or not functioning effectively, a leader seeks to understand the forces in the background that shape and move the individuals who are in the foreground.

3. Groups are not particularly rational entities. More often than not they are inherently irrational — that is, all manner of unconscious sentiments are swirling beneath the surface that impact the flow of the group dynamic and the people's capacity to ad-

dress problematic realities. These unconscious sentiments generate work avoidance mechanisms such as
scapegoating deviants, attack on the authority, dis-
placement of responsibility, and the generation of decoy issues and false tasks.

4. Leadership interventions must address not just
the individual but the system and the factional dynamics of the system if progress is to be made and sustained. The preoccupation with personalities and individual psychology without diagnosing the systemic
dynamics is incomplete and limited in usefulness.

5. Leaders must consider *what part* of the system
needs to make adjustments in values or perspectives
to deal with the threat, danger, or opportunity. But
in addressing the part of the system that is deficient,
they must also consider interdependence of the various parts and how each part impacts the whole.

6. Individual analysis as it pertains to an adaptive
challenge should be about the roles people play in the
system. People play roles and are generally unconscious of the fact they are playing a role. A role is a
group construct.

7. A role is a consistent and persistent pattern of
behavior and response that serves a particular purpose for the group. We may consciously try to change
our role and succeed, but that is an adaptive challenge
in and of itself. But the power of a system is such that
it can hold us in a particular place, or use us in a particular way, so that we have little room to maneuver.
In other words, there are patterns and dynamics in
the relationship between the individual and the group
that reinforce certain behaviors that make it difficult
for individuals to change, even should they so desire.
While roles can be constraining, they can also be a resource that allows us to make interventions or support
others in the doing of adaptive work.

117

8. Anyone attempting to exercise leadership should seriously reflect on whether their role is being used by the group for work avoidance, distancing from responsibility, feeding group hungers, and putting a false set of tasks before the group — rather than engaging the difficult work of progress.

9. It is important to distinguish between *self* and *role*. A role is something that we take up or have imposed upon us. But the role is not us. The moment one collapses the role with self, the ability to think clearly and exercise leadership is significantly reduced. Often, when people attack the leader or get upset with the boss or one becomes a lightning rod for a particular issue, it is because that person embodies the issue for the group. It is important not to take the attack personally. Likewise, when the group applauds and adores the leader, one should not be seduced or succumb to the excessive longings of the people to solve their problems for them and produce the magic solution.

Authority

1. Authority is a very ancient and important orienting function for any group, be it a family, school, club, community, or nation. But the authority role is difficult when competing factions have varied expectations and hungers that they demand the authority figure to respond to or sate.

2. Some of the basic expectations of any group include the provision of means to obtain fundamental necessities, protect boundaries, control conflict, maintain the norms for the group, and advance its interests.

3. Groups also look to the authority figure to embody the "mantle of the ideal." Such figures are ex-

pected to display the values and practices that are meaningful to the people.

4. Failure to fulfill people's expectations, feed enough of their hungers, or be the mantle of the ideal can lead to the de-authorization of the authority figure.

5. Authority figures are often used by the people to avoid confronting the real work of progress. Some groups constantly replace their authority figures, but things do not get any better. The authority figure simply is an extension of themselves.

6. The ways people use authority to avoid real work include: 1) the tendency to want the authority to be a magician and fix everything; 2) excessive dependence on or deference to the authority with the hope that the authority will make their life easier or bestow benefits (status and rewards) on them for their loyalty; and, 3) unnecessarily, even irrationally, fighting with or undermining the authority because the person is seen as an impediment or a threat by virtue of their position or dominance.

7. While authority can be a massive work-avoidance issue for a group, its function can also be used productively to promote adaptive work. Authority should be used to formulate a holding environment to contain the group's work-avoidance patterns, which are inevitable when dealing with tough problems. A holding environment consists of the set of values, rules, norms and boundaries that hold the group in their conflicts and problem-solving pursuits.

8. While a tight holding environment can promote group identity, it can also limit the creativity needed to produce adaptive solutions. Authority plays an important role in strengthening or loosening the holding environment depending on what challenge the people

face and the amount of creativity and conflict that is needed to produce a solution.

9. Because people look to authorities for guidance and answers, it is very seductive for authorities to act like they know where they are going when they do not and to feed the people's hungers when they should be feeding themselves. This can be dangerous, as it promotes dependence and delusions. The challenge of authority figures is to use their authority to direct people's attention to aspects of reality that people are avoiding or refuse to contend with, and call attention to the contradiction in values between what people espouse and how they behave, even when it is painful for the people.

10. Authority often serves as a repository for people's hopes, fears, aspirations, and pain. This can be a heavy burden for the authority figure. The authority figure, in the exercise of leadership, must seek to give enough of the burden of responsibility back to the people at a rate they can tolerate.

11. Authority can also promote learning and adaptive work by protecting the deviant voices in the system that others seek to block.

Intervention.

1. The essence of leadership is in intervention to direct attention to a problematic concern. This is always difficult, as people have so many other concerns competing for their attention. Leadership interventions must be creative in order to get and hold attention.

2. A creative intervention necessitates stepping beyond the normal way of speaking and listening and ensuring that timing, pacing, voice modulation, and partnering are features of the intervention process.

3. A leadership intervention is a "perturbing force" that stirs the group to confront an issue that they have been avoiding or to pursue a course of action that they have been reluctant to pursue.

4. Perturbing can be done through provoking and evoking. A provocative intervention stirs the group into action by generating dissonance. An evocative intervention stirs people to action by appealing to higher values or noble sentiment. Excessive provocation can lead to rebellion, and excessive evoking can generate group dependency. Leadership interventions should use multiple means to stir the group. Doing the same thing all the time will reduce the potency of the interventions.

5. In order to intervene, one must be able to understand the difference between holding steady and holding back. Holding steady is being in a state of engagement and watching to see how the group is responding, while holding back is a defense mechanism that distances oneself from responsibility for contributing.

6. Successful interventions need partners to help orchestrate the intervention. It is dangerous to lead alone. Partners help to keep the attention on the problem. They clarify issues. They highlight the leader's blind spots and provide feedback with regard to what is working or not working.

7. A leader must develop the capacity to make spontaneous and improvised interventions, and planned, designed, and highly strategic interventions. All are needed, since leadership interventions must deal with the long-term challenges of mobilizing people to do adaptive work and the day-to-day challenges of sustaining attention, maintaining levels of engagement, and minimizing distractions. Adaptive work by its

very nature unfolds in a dynamic and unpredictable manner.

8. Not all adaptive problems are clearly framed or obvious to the people. Therefore, interventions should be used to ripen issues so that the problem steadily builds a constituency for engagement.

Creativity and Leadership.

1. It might be said that there can be no leadership without creativity. Generally there can be no adaptive work without creative problem-solving. Creativity is the engine of progress. Adaptive work necessitates pushing boundaries and frontiers through creative exploration. Creativity is needed because the solution for a difficult adaptive challenge may lie outside the current repertoire of responses.

2. Thomas Kuhn, in *The Structure of Scientific Revolutions*, says that group knowledge exists within a paradigm—a shared set of myths, beliefs, and assumptions. Leadership prods and pokes at the myths, beliefs, and assumptions of the group to see if the paradigm should be broadened, modified, or abandoned.

3. Creative leadership requires the courage to challenge a prevailing myth, norm, practice, or value. It requires courage because challenging the group might be threatening to a group that is rigidly attached to its paradigm, a source of comfort and security to the group.

4. Creative adaptations in the group, organization, or community generally occur in a state of tension or conflict. Rollo May argues that conflict is an inherent part of the creative process and generates the sparks that ignite the consideration and exploration of alternatives and unconsidered propositions.

5. The conflict associated with the creative process is the product of competing values and perspectives rubbing up (even clashing) against each other. Through this engagement process, the spark of new insights and possibilities can emerge. Authority, however, must manage the disequilibrium associated with the creative process to ensure that it does not turn destructive. Indeed, creative work can be terrifying for people and lead them to flee, scapegoat a faction, or even kill the leader. Rollo May says that creativity provokes the jealousy of the gods!

Assassination and the Challenge of Staying Alive.

1. Leadership can be dangerous, as you are often on the razor's edge or in a vulnerable space. Groups neutralize leaders and authority figures who challenge prevailing wisdom by assassinating them, undermining them, marginalizing them, or overthrowing them. Dissident voices, who provide some leadership without any authority and challenge the group by calling attention to the group's hypocrisy, can also be attacked, marginalized, or silenced.

2. All groups have mechanisms for neutralizing the provocative voice or the unacceptable authority figure.

3. Dissident and provocative voices that are moving the group toward important aspects of reality should be protected and partnered with to ensure that they are not killed off, and that the issue they represent remains alive.

4. The leadership challenge is to keep the issue on the table and yourself off the table. You want people dissecting the issue and not dissecting you.

5. Do not be a martyr or encourage martyrdom in others. Martyrs are often used by groups to avoid

the real work. They perpetuate an "us versus them" dynamic, encourage delusionary and grandiose thinking, and promote counterproductive battles that persist.

6. To minimize the risk of assassination, have partners and confidants to protect you. Also, pace the work. Do not become a self-righteous crusader who seeks to impose your plan upon others. While a group can hold terrible views and engage in negative practices, it is necessary for the leader to have a degree of sensitivity and understanding as to how those views and practices came into existence.

7. It is important for those in the role of lightning rod to distinguish between attacks that are acts of sabotage or neutralization and the attacks that are inevitable on an important issue with which people are struggling. As a lightning rod, one must be able to hold steady and ascertain the deeper concerns and fears that are underlying the attack.

Purpose, Task and Work Avoidance.

1. Leaders need a compelling sense of purpose to hold them as they pursue the exercise of leadership.

2. Purpose is a place to *come from* rather than a destination to get to. It shapes your relationship to the problem. It generates a *way of being* with the problem.

3. Purpose serves as a holding environment for the leader and the group when doing adaptive work. It holds your doubts and your aspirations. It keeps you in the game. It gives you reason to intervene and challenge. It also gives you reason to learn and be curious.

4. Purpose must be self-generated and not imposed. There is no purpose beyond what you or the people generate.

5. Maintaining a sense of purpose is essential to keep the people from fleeing or succumbing to their hungers. Purpose puts something at stake. It puts fire in the belly.

6. All groups have work-avoidance mechanisms that they employ to distance themselves from responsibility for their real problems—this is because their sense of purpose is very weak, insufficiently constructed, or connected to the perpetuation of their identity.

7. Excessive work avoidance might produce a state of purposelessness or narrowly defined self-interest that leads people to act irresponsibly and endanger themselves and their community. The goal of leadership is to work with the group to generate a sense of purpose and get connected to a worthwhile and noble aspiration—not necessarily a vision (a place to get to) but an aspiration (something to live and work for).

8. Work-avoidance mechanisms include any group activity that leads people away from the engagement of a problematic reality—such as fighting; competing for resources or status; succumbing to one's hungers; politicking; fleeing from responsibility; disengagement; the generation of false visions and delusions; and the pursuit of false tasks that seem momentarily attractive but really have nothing to do with sustainable progress.

9. A leader must be willing to tolerate a degree of work avoidance in the system (it is inevitable), in the same way that wise parents will tolerate a degree of work avoidance in their teenage child as the child engages in the developmental process of transitioning into adulthood. No group goes straight to the problem, but generally it goes around in circles for some time. This can be an important learning process that occurs in a state of disequilibrium and allows for discoveries to be made.

10. It is important to distinguish exercising leadership with a sense of purpose from crusading. Crusading is the thoughtless pursuit of a goal. Crusaders, although extremely committed, can do tremendous damage. Rather than engaging in learning and maintaining an attitude of curiosity, they hold onto a dogma as a truth. Leadership promotes thinking, a thinking that allows for more effective action in the face of complex problems.

Listening and Leadership.

1. If one cannot listen, one cannot exercise leadership. So much of leadership work is about listening.

2. Listening allows one to access the sentiment of the group and its respective factions, and to hear the people's fear, pain, and despair — and their hopes, dreams, and desires.

3. There is no such thing as empty listening. Our listening is heavily warped and distorted with assessments and judgments as we listen from a particular position or from the tuning of our harp strings (the significant voices in our heads).

4. To exercise successful leadership effectively, one must have a degree of self-knowledge or a willingness to pursue such knowledge. In order to build self-knowledge, one must inquire into how one listens to access reality. Consider the following questions:

a. Where do I listen from (my assumptions, my positions, my loyalties)?

b. Whom do I listen to? (What people or factions do I consider worthwhile, and whom do I discount? Do I listen to the voices on the margins? Can I really listen to my enemy?)

c. What do I listen for? (We often listen for agreement, acknowledgment, acceptance, or to feed our hungers, whatever they might be. In the exercise of leadership one has to listen for "what's missing" in the group, the sentiment of the group, and the music beneath the words).

5. The challenge of leadership is to deepen and broaden your listening capacity. In the class, the music occasionally played is used as a metaphor for the challenge of listening to sounds that do not fit neatly into our paradigm. How difficult it is to give undivided attention and listen and hear the meaning and sentiment that is imbedded in the sounds, as quaint as they might be.

6. The challenge of leadership is to also to listen to the deviant voice, the individual or faction that is trying to raise an issue or concern that is in opposition to the prevailing view. A deviant can easily be killed off and marginalized. Authority figures should protect the deviant voice, and the leadership work is to tackle the concern the deviant is raising and not to tackle the deviant.

7. The challenge of leadership is to listen to your enemies — to really hear their concerns and what is driving their opposition.

8. The challenge of leadership is to listen to voices on the margins. Who is being pushed to the side and why? What perspectives do they embody that the larger group is having difficulty incorporating?

9. The challenge of leadership is to listen to the frontline. Can you really hear the concerns, fears, advice, and perspectives of those on the frontline who are fighting the battles, interacting with customers, or doing the real implementation work?

10. The challenge of leadership is to listen to confidants and allies who can talk straight and tell you when you are being a jerk or causing an unnecessary mess.

Inspiration and Leadership.

1. Groups need inspiration to face demanding challenges. Inspiration is about breathing life into the group. Inspiration is needed to keep people engaged in the work, particularly when one is tired and finds it difficult to persist. Inspiration can serve to "keep the fire burning" and to keep people focused on the work that matters, particularly when the people are tired and seek to flee.

2. It is difficult to be inspirational when one is being excessively technical about the work or stuck in a particular role.

3. Inspiration can be used to shift people's relationship to an issue, concern, or challenge. It can help them see what is at stake in a way that connects their own values to noble aspirations.

4. Inspiration serves to remind people of what is possible. Inspiration speaks to higher sentiments, the "divine spark" and the "noble" side of people. The dark side is often exploited by counterfeit leaders to stir people into a state of vengeance or to promote a false sense of security.

5. Leaders must take responsibility for how they employ and deploy inspiration. Inspiration can be a dangerous tool. It can be used to perpetuate work avoidance in a social system by generating false hope, by putting faith in decoy issues, or by inspiring people to scapegoat others and to avoid taking responsibility for their condition.

6. Inspiration can also lead to excessive dependency on the authority figure or one exercising leadership. The people might call the leader charismatic and project onto him or her some magical quality, thereby

diminishing their own capacity and responsibility for doing adaptive work and exercising creativity in the face of difficult problems.

7. According to the Odin myth, the gift of poetic inspiration was the most important gift of the gods, and therefore the dwarfs and giants were constantly fighting to steal the sacred mead (potion) that gave access to the gift. Whoever had the mead had the ultimate power over the people to do with them as they wanted! The abuse of the mead, as we see in the Odin myth, may lead to the destruction of all.

8. Inspiration must be connected to aspiration. Inspiration breathes life into the group, while aspiration is breathing out, or taking steps toward the desired purpose.

9. To provide genuine inspiration, the leader must tap into one's personal bank of pain, joy, and experience and connect it to the reality of where the people are. In other words, there must be an authentic connection between the leader and the group, a connection that transcends the despair of the moment and creates a window of possibility for the group.

Personal Work of Leadership.

1. To exercise leadership, one must be like the Norse god Odin and have a process for increasing in wisdom and understanding in order to minimize the damage your power might cause and maximize the effectiveness of your interventions.

2. Use partners to detect what is missing due to your blind spots. A leader must ask, "How am I wrapped up in this problem in ways that I don't see?" Partners can help answer that question.

3. Detect when you are leading by crusading. It is easy to become excessively passionate and single-

minded about the rightness of your pursuit, failing to see other options and alternatives and to make adjustments in your assumptions, strategies, and actions. Crusaders become obnoxious about their cause and provide no room for serious adaptive work.

4. Detect when your narcissism is getting in the way. Many people who enter the political arena or seek to lead with authority have narcissistic tendencies. This is not necessarily bad. Narcissism is the feeling of self-importance and the exaggeration of your value and contribution. Leaders must always be watching to ensure their narcissism does not get in the way of the real work. Excessive narcissism produces counterfeit leadership because the spotlight falls on the person and not the problem.

5. Detect when your convictions are getting in the way of your capacity for accurate diagnosis. We all have beliefs, desires, and convictions that frame our view of the world. In doing thorough diagnostic work, it may be necessary to put our convictions to the side in order to listen, observe and distinguish what really is going on.

6. Detect when your hungers are getting in the way. Hungers are desires that we seek to fulfill in the course of our normal pursuits. It is easy for our hungers to distract us from exercising leadership and thereby sabotage our capacity to make a serious contribution. Hungers include the desire for status, prominence, dominance, control, territory, acceptance, adoration, etc.

7. Detect when your factional loyalties are impeding the doing of adaptive work. We all have attachments to family, friends, communities, and nations. Our loyalties to our respective factions can often lead us to not challenge our own group or to disappoint

them. If our attachment is too strong, we may promote tribalism at the expense of progress.

8. Detect the people's reality; stay grounded, and wander. You must be able to find out what is really going on in the system. This will require taking off the robes of authority, leaving one's hallowed office, and going out among the people to sense for yourself what is really happening without any filter.

9. Detect when to move to the side or get out of the way. Sometimes the work of leadership is to get out of the way and go home. In other words, there is nothing more that you can do, and if you persist in staying connected to the challenge, your presence will impede progress. It might be time to get lost and let someone else take over.

APPENDIX II

U.S. ARMY OFFICER EVALUATION REPORT

OFFICER EVALUATION REPORT For use of this form, see AR 623-3 the proponent agency is DCS, G-1.					FOR OFFICIAL USE ONLY (FOUO) SEE PRIVACY ACT STATEMENT IN AR 623-3	

PART I - ADMINISTRATIVE DATA

a. NAME (Last, First, Middle Initial)	b. SSN	c. RANK	d. DATE OF RANK (YYYYMMDD)	e. BRANCH	DESIGNATED SPECIALTIES / PMOS (WG)

g.1. UNIT, ORG., STATION, ZIP CODE OR APO, MAJOR COMMAND	g.2. STATUS CODE	h. REASON FOR SUBMISSION

i. PERIOD COVERED		j. RATED MONTHS	k. NONRATED CODES	l. NO. OF ENCL	m. RATED OFFICER'S AKO EMAIL ADDRESS (.gov or mil)	n. UIC	o. CMD CODE	p. PSB CODE
FROM (YYYYMMDD)	THRU (YYYYMMDD)							

PART II - AUTHENTICATION (Rated officer's signature verifies officer has seen completed OER Parts I-VII and the admin data is correct)

a. NAME OF RATER (Last, First, MI)	SSN	RANK	POSITION	SIGNATURE	DATE (YYYYMMDD)
b. NAME OF INTERMEDIATE RATER (Last, First, MI)	SSN	RANK	POSITION	SIGNATURE	DATE (YYYYMMDD)
c. NAME OF SENIOR RATER (Last, First, MI)	SSN	RANK	POSITION	SIGNATURE	DATE (YYYYMMDD)

SENIOR RATER'S ORGANIZATION	BRANCH	SENIOR RATER TELEPHONE NUMBER	E-MAIL ADDRESS (.gov or .mil)	
	d. This is a referred report, do you wish to make comments? ☐ Yes, comments are attached ☐ No		e. SIGNATURE OF RATED OFFICER	DATE (YYYYMMDD)

PART III - DUTY DESCRIPTION

a. PRINCIPAL DUTY TITLE	b. POSITION AOC/BR

c. SIGNIFICANT DUTIES AND RESPONSIBILITIES. REFER TO PART IVa, DA FORM 67-9-1.

PART IV - PERFORMANCE EVALUATION - PROFESSIONALISM (Rater)

CHARACTER Disposition of the leader: combination of values, attributes, and skills affecting leader actions

a. ARMY VALUES (Comments mandatory for all "NO" entries. Use PART Vb.)	Yes No		Yes No
1. HONOR: Adherence to the Army's publicly declared code of values		5. RESPECT: Promotes dignity, consideration, fairness, & EO	
2. INTEGRITY: Possesses high personal moral standards; honest in word and deed		6. SELFLESS-SERVICE: Places Army priorities before self	
3. COURAGE: Manifests physical and moral bravery		7. DUTY: Fulfills professional, legal, and moral obligations	
4. LOYALTY: Bears true faith and allegiance to the U.S. Constitution, the Army, the unit, and the soldier			

b. LEADER ATTRIBUTES / SKILLS / ACTIONS: First, mark "YES" or "NO" for each block. Second, choose a total of six that best describe the rated officer. Select one from ATTRIBUTES, two from SKILLS (Competence), and three from ACTIONS (LEADERSHIP). Place an "X" in the appropriate numbered box with optional comments in PART Vb.
Comments are mandatory in Part Vb for all "No" entries.

b.1. ATTRIBUTES (Select 1) Fundamental qualities and characteristics	1. MENTAL YES NO Possesses desire, will, initiative, and discipline	2. PHYSICAL YES NO Maintains appropriate level of physical fitness and military bearing	3. EMOTIONAL YES NO Displays self-control, calm under pressure
b.2 SKILLS (Competence) (Select 2) Skill development is part of self-development; prerequisite to action	1. CONCEPTUAL YES NO Demonstrates sound judgment, critical/creative thinking, moral reasoning	2. INTERPERSONAL YES NO Shows skill with people: coaching, teaching, counseling, motivating and empowering	3. TECHNICAL YES NO Possesses the necessary expertise to accomplish all tasks and functions
	4. TACTICAL Demonstrates proficiency in required professional knowledge, judgment, and warfighting		YES NO

b.3. ACTIONS (LEADERSHIP) (Select 3) Major activities leaders perform: influencing, operating, and improving

INFLUENCING Method of reaching goals while operating / improving	1. COMMUNICATING YES NO Displays good oral, written, and listening skills for individuals / groups	2. DECISION-MAKING YES NO Employs sound judgment, logical reasoning and uses resources wisely	3. MOTIVATING YES NO Inspires, motivates, and guides others toward mission accomplishment
OPERATING Short-term mission accomplishment	4. PLANNING YES NO Develops detailed, executable plans that are feasible, acceptable, and suitable	5. EXECUTING YES NO Shows tactical proficiency, meets mission standards, and takes care of people/resources	6. ASSESSING YES NO Uses after-action and evaluation tools to facilitate consistent improvement
IMPROVING Long-term improvement in the Army, its people and organizations	7. DEVELOPING YES NO Invests adequate time and effort to develop individual subordinates as leaders	8. BUILDING YES NO Spends time and resources improving teams, groups and units; fosters ethical climate	9. LEARNING YES NO Seeks self-improvement and organizational growth; envisioning, adapting and leading change

c. APFT:	DATE:	HEIGHT:	WEIGHT:			
d. OFFICER DEVELOPMENT - MANDATORY YES OR NO ENTRY FOR RATERS OF CPTs, LTs, CW2s, AND WO1s. WERE DEVELOPMENTAL TASKS RECORDED ON DA FORM 67-9-1a AND QUARTERLY FOLLOW-UP COUNSELINGS CONDUCTED?				YES	NO	NA

DA FORM 67-9, OCT 2011	PREVIOUS EDITIONS ARE OBSOLETE.	Page 1 of 2 APD PE v1.00ES

NAME	SSN	PERIOD COVERED

PART V - PERFORMANCE AND POTENTIAL EVALUATION *(Rater)*

a. EVALUATE THE RATED OFFICER'S PERFORMANCE DURING THE RATING PERIOD AND HIS/HER POTENTIAL FOR PROMOTION

☐ OUTSTANDING PERFORMANCE, MUST PROMOTE ☐ SATISFACTORY PERFORMANCE, PROMOTE ☐ UNSATISFACTORY PERFORMANCE, DO NOT PROMOTE ☐ OTHER *(Explain)*

b. COMMENT ON SPECIFIC ASPECTS OF THE PERFORMANCE, REFER TO PART III, DA FORM 67-9 AND PART IVa, b, AND PART Vb, DA FORM 67-9-1.

c. COMMENT ON POTENTIAL FOR PROMOTION.

d. IDENTIFY ANY UNIQUE PROFESSIONAL SKILLS OR AREAS OF EXPERTISE OF VALUE TO THE ARMY THAT THIS OFFICER POSSESSES. FOR ARMY COMPETITIVE CATEGORY CPT ALSO INDICATE A POTENTIAL CAREER FIELD FOR FUTURE SERVICE.

PART VI - INTERMEDIATE RATER

PART VII - SENIOR RATER

a. EVALUATE THE RATED OFFICER'S PROMOTION POTENTIAL TO THE NEXT HIGHER GRADE

I currently senior rate _____ officer(s) in this grade

A completed DA Form 67-9-1 was received with this report and considered in my evaluation and review ☐ YES ☐ NO *(Explain in c)*

☐ BEST QUALIFIED ☐ FULLY QUALIFIED ☐ DO NOT PROMOTE ☐ OTHER *(Explain below)*

b. POTENTIAL COMPARED WITH OFFICERS SENIOR RATED IN SAME GRADE (OVERPRINTED BY DA)

c. COMMENT ON PERFORMANCE/POTENTIAL

☐ ABOVE CENTER OF MASS
(Less than 50% in top box; Center of Mass if 50% or more in top box)

☐ CENTER OF MASS

☐ BELOW CENTER OF MASS RETAIN

☐ BELOW CENTER OF MASS DO NOT RETAIN

d. LIST THREE FUTURE ASSIGNMENTS FOR WHICH THIS OFFICER IS BEST SUITED. FOR ARMY COMPETITIVE CATEGORY CPT, ALSO INDICATE A POTENTIAL CAREER FIELD FOR FUTURE SERVICE.

APPENDIX III

RECOMMENDED CHANGES
TO LEADER CORE COMPETENCIES
OUTLINED IN THE
ARMY LEADER DEVELOPMENT STRATEGY,
NOVEMBER 2009

Leader Core Competencies extract from the Army Leader Development Strategy (ALDS), with recommended additions.

LEADER CORE COMPETENCIES

Army leaders apply their character, presence, and intellect in leading our nation's Soldiers. The expectations for what leaders should do regardless of the situation are captured in the Army's core leader competencies. Core leader competencies are defined as groups of related behaviors that lead to successful performance common throughout the organization and are consistent with the organization's values. There are eight leader competencies that fall into four areas:

One who leads. Provides vision through purpose, motivation, universal respect, and direction to guide others. Extends one's influence beyond the chain of command to build partnerships and alliances to accomplish complex work. Leading is conveyed by communicating (imparting ideas) and setting the example.

One who develops. Leads organizations by creating and maintaining a positive environment and by investing effort in their broadening, and that of others, to achieve depth and breadth. Developing includes assessing needs to improve self, others, and the organization.

One who achieves. Focuses on what needs to be accomplished. Has an expeditionary mindset and can adapt to unanticipated, changing, and uncertain situations. Achieving in the short term is about getting results but in the long term, it is about setting the vision to obtain objectives.

One who mobilizes. Navigates organizations through adaptive challenges by confronting people with the reality of changed conditions. Provides a learning environment for the group to discover and develop adaptive solutions.

U.S. ARMY WAR COLLEGE

Major General Gregg F. Martin
Commandant

*****

STRATEGIC STUDIES INSTITUTE

Director
Professor Douglas C. Lovelace, Jr.

Director of Research
Dr. Antulio J. Echevarria II

Author
John B. Richardson IV

Director of Publications
Dr. James G. Pierce

Publications Assistant
Ms. Rita A. Rummel

*****

Composition
Mrs. Jennifer E. Nevil

www.ingramcontent.com/pod-product-compliance
Lightning Source LLC
Chambersburg PA
CBHW081408270326
41931CB00016B/3413